Aspects of Oral Molecular Biology

# Frontiers of Oral Physiology

Vol. 8

Series Editor
*D.B. Ferguson*, Manchester

KARGER

Basel · München · Paris · London · New York · New Delhi · Bangkok · Singapore · Tokyo · Sydney

# Aspects of Oral Molecular Biology

Volume Editor
*D.B. Ferguson,* Manchester

24 figures and 6 tables, 1991

KARGER

Basel · München · Paris · London · New York · New Delhi · Bangkok · Singapore · Tokyo · Sydney

# Frontiers of Oral Physiology

Library of Congress Cataloging-in-Publication Data
Aspects of oral molecular biology / volume editor, D.B. Ferguson.
p. cm. – (Frontiers of oral physiology; vol. 8)
Includes bibliographical references and index.
1. Mouth – Physiology. 2. Molecular biology. 3. Teeth – Growth – Molecular aspects.
4. Salivary proteins – Synthesis – Regulation.
I. Ferguson, D.B. II. Series.
[DNLM: 1. Molecular Biology. 2. Mouth — physiology. 3. Mouth – ultrastructure.
4. Tooth – physiology].
ISBN 3-8055-5261-0

Drug Dosage

The authors and the publisher have exerted every effort to ensure that drug selection and dosage set forth in this text are in accord with current recommendations and practice at the time of publication. However, in view of ongoing research, changes in government regulations, and the constant flow of information relating to drug therapy and drug reactions, the reader is urged to check the package insert for each drug for any change in indications and dosage and for added warnings and precautions. This is particularly important when the recommended agent is a new and/or infrequently employed drug.


ISBN 3–8055–5261–0

# Contents

# Foreword

## The Present State of Molecular Biology Studies in Relation to the Oral Region

The tremendous advances which have taken place in the understanding of the complex molecular events involved in the development and continuing metabolism of tissues, particularly in relation to the transmission of information from genetic material, have opened up new fields of research in all the biomedical sciences. Oral biology is no exception to this. I began planning this volume some 2 years ago; the delay in its appearance springs from the many refusals of my invitations to contribute to it. These refusals had a common theme: the volume was too early because the research was still in progress and much work remained to be done. The tone of these letters confirmed my view that molecular biology was the subject of choice for a volume in a series with the word frontiers in its title. There is a risk that the contents of this volume will rapidly become dated because of their very topicality, but the quality of the contributors makes that less likely.

The techniques and discoveries of molecular biology have developed with remarkable speed and there now exists a situation in which undergraduates are being taught facts and methods outside the repertoire of many experienced research workers in the general field of oral biology. This book is intended in part to improve that situation. The chapters contributed by Dr. Russell and myself are written partly with that in mind. If this is insufficient the reader will find it worthwhile to consult one of the excellent textbooks of molecular biology now available – the second edition of 'Molecular Biology of the Cell', by Alberts, Bray, Lewis, Raff, Roberts and Watson, published by Garland Publishing Inc., New York and London, 1989, is an excellent example.

The term 'molecular biology' or 'molecular cell biology' is now used to describe a subject area which includes traditional biochemistry, cell ultrastructure and physiology (cell biology), and genetics. Within these broader areas there has been a concentration on molecular genetics, dealing with the transcription of genetic information, the translation into protein molecules and the post-translational modification of these proteins, and on developmental signals such as growth factors and the molecules associated with tissue interactions during development. Many of the techniques of molecular biology are those of traditional biochemistry. To these have been added immunological methods for identification of molecules. However, the techniques which are more specific to molecular biology are those which have been developed for working with DNA and RNA, and in particular the techniques for cloning these molecules in host bacterial cells.

The chapters of this book have been arranged according to the tissues which have been studied – first the dentition, with specific chapters on enamel and dentine, then the oral bacteria, and finally the salivary components, enzymes, proline-rich proteins and glycoproteins. In this overview I shall consider them in terms of technique. In considering the genetic control of the dentition, Dr. Lesot ranges widely over inheritance studies, identification of genes linked to the development of the dentition and tooth types, cell-matrix and cell-cell interactions, and the localization of specific proteins such as epidermal growth factor, fibronectin, cadherins, actin and talin. All these may be implicated in the development of a dentition correctly situated in the jaw, with the different tooth types correctly arranged.

The three chapters on the dental tissues concentrate on molecules specifically associated with their development. Drs. Termine and Young give an account of the differences between the enamelins and the amelogenins, and describe how primary amino acid sequences have been established for the amelogenins and the products of bovine ameloblast mRNA have been identified. Drs. Herold and Rosenbloom extend these observations by describing the localization of amelogenins and enamelins in mammalian and nonmammalian species, using monoclonal antibodies to bovine proteins. Dr. Thesleff and her co-workers have explored the molecular basis of the differentiation of the dental mesenchyme which will eventually produce dentine. They have looked at the expression of a number of key molecules in the developing mesenchyme-binding sites for epidermal growth factor (EGF), the matrix protein tenascin, a cell-surface proteoglycan, and a protein found in high concentration in the dental papilla at the bell stage of development. Tenascin and the cell-surface proteoglycan seem to result from the interaction of the dental mesenchyme with epithelium.

Dr. Russell's contribution on the impact of molecular genetic methods on oral microbiology moves the focus away from developmental studies to an area where the application of new techniques based on DNA and RNA analysis and cloning has begun to answer some previously insoluble problems. Thus, studies of chromosomal DNA have facilitated species identification and enabled the relationships between oral bacterial species to be determined. Genetic probes have been developed to identify species. Cloning of genes from streptococci has been used to study their cell surface proteins and to examine their metabolism, particularly in relation to enzymes such as the glucosyl transferases.

The sequencing of proteins and identification and cloning of genes for these proteins has transformed the world of salivary research also. Extensive studies of inheritance, structure and genetic coding of salivary proteins have been developing over some 15 years. Dr. Karn ranges over this expanding field of knowledge with a chapter on salivary enzymes. Salivary amylase has been sequenced and its genes identified in human and rodent tissue. Both salivary and pancreatic amylase have been studied and the human genome is now known to contain three salivary and two pancreatic amylase genes together with two pseudogenes, all situated on human chromosome 1 (1p21). In the mouse they lie on chromosome 3. The genetics of other salivary proteins – in particular the proline-rich proteins (further described in the next chapter), statherin, histatin, cystatin and the mouse androgen-binding protein – have also been explored. In another chapter (chapter 7) I give an account of present knowledge of the salivary proline-rich proteins. This group of closely similar proteins probably comprises almost all the proteins of saliva apart from those with enzymatic activity and the more complex glycoproteins usually termed mucins. They are virtually unique to saliva. The major proline-rich proteins have been sequenced and shown to consist of a small number of repeat sequences of amino acids with very small variations between the different proteins. Their genes have been identified and located on chromosome 12 (12p13.2) in humans. Finally, Dr. Tabak contributes a chapter on the complex and interesting field of salivary mucin research. Progress here has been slower because, as he points out, the very properties which make these glycoproteins of value in protecting the oral mucosa also make them very resistant to analysis. Despite this, the routes by which these glycoproteins are synthesized are being delineated and the post-translational processing mechanisms are becoming better understood. The effect of different stimuli on glycoprotein synthesis has been examined. However, the fundamental genetic control of mucin formation is still being explored.

It will be apparent from the reviews in this book that much progress has been made in the study of the molecular events which occur in the development of the dental tissues and the dentition as a whole, in the characterization of oral microorganisms by their genetic fingerprint, and in the study of salivary protein genetics, composition and synthesis. There are still many unsolved problems but, if the state of knowledge revealed here is compared with that of only 5 years ago, the pace of discovery is such that it seems likely that their number will be greatly reduced within the next few years.

*D.B. Ferguson*

Ferguson DB (ed): Aspects of Oral Molecular Biology.
Front Oral Physiol. Basel, Karger, 1991, vol 8, pp 1–16

# Genetic Control of Dentition

*Hervé Lesot*[1]

Institut Biologie Médicale, INSERM CJF 88–08, Faculté de Médecine,
Strasbourg, France

Dentition, which is characterized by the number, localization and shape of teeth, shows variations between species and within species. Heterodonty appears early in mammalian evolution and demonstrates the existence of four tooth classes in most mammals. These four classes do not correspond to clearly distinct compartments since evolutionary processes may have led to progressive changes in tooth shape and size, and thus could have resulted in molarization of premolars, retention of deciduous teeth, premolarization of canine, caninization of premolar or change in the number of molars [1].

Dentition results from complex morphogenetic processes leading to the sequential appearance of a series of teeth and to the coordinated development of the maxillary and mandibular constituents. During tooth development, the genetic control of the dental matrix, and especially the enamel (see other chapters in this volume) is very rigid, whilst the control of tooth form and volume is more plastic and involves both major genes and polygenic inheritances as well as environmental factors [2].

The aims of this review are, firstly, briefly to summarize data concerning the genetic influences on different parameters of the dentition and, secondly, to discuss developmental aspects allowing experimental approaches to the molecular mechanisms, genetic or epigenetic, which regulate odontogenesis.

[1] The author gratefully acknowledges Dr. J.V. Ruch for stimulating discussions and comments.

## *Genetic Aspects*

### *Tooth Number*

Congenital variations in the number of teeth, anodontia, hypodontia and hyperdontia, are relatively common malformations that appear to result from different etiological mechanisms not yet clearly understood. *Anodontia*, absence of all teeth, is a rare disorder whose genetic basis is unknown [3]. *Hypodontia* results from the agenesis of one or more teeth. In most cases, hypodontia has no apparent environmental etiology and might result from a single gene defect [4], often transmitted as an autosomal-dominant trait with incomplete penetrance and variable expressivity [3, 5]. The incomplete penetrance suggests the presence of a mechanism whereby epistatic genes interfere with the phenotypic expression of the hypodontia gene [3]. *Hyperdontia* is much less common than hypodontia. It can be observed in cases of Gardner's syndrome, or in cases of cleidocranial dysplasia both of which are autosomal-dominant disorders [6].

### *Tooth Shape and Size*

The number of postmitotic functional odontoblasts and ameloblasts determine the size of tooth crown and the spatial distribution of these cells condition the shape of the crowns [7]. However, most studies concerning changes in tooth shape and size do not distinguish between variations in the thickness of enamel and dentine and modifications in the number of dental cells.

The question of tooth shape is linked to that of tooth class. However, even within a single class, differences can be observed between teeth. Cheek teeth, in many species, can be ranged in a gradient from a simple tooth to a complex one, depending on cuspal morphology [8]. Isolated mouse tooth germs in culture develop their normal cusp patterns from a very early stage, showing that cusp development is an intrinsic process.

The dental morphology in patients with Turner's syndrome, when compared with that of relatives, shows a less shovel-shaped lateral incisor, a shallower fossa on the central incisor and a reduction of the distolingual cusp [9]. These observations suggested that a size-independent morphological reduction of tooth crowns is associated with the sex chromosome aberration of 45,X females [9].

Studies of the genetic influences on the size of tooth crowns in humans have demonstrated that genes present on both X and Y chromosomes were involved [10]. Males with Klinefelter's syndrome tend to have larger permanent teeth than normal, as do 47,XYY males, whereas permanent tooth size is reduced in Turner's syndrome [11–13]. Tooth size appears to be a polygenic autosomal trait, influenced by the sex chromosome complement. The X and Y chromosomes both seem to have a similar influence on enamel thickness, possibly by regulating amelogenesis, whereas the Y chromosome appears to have a greater effect on dentine thickness, perhaps by modifying cell proliferation [14]. The examination of two males with deletions of the long arm of the Y chromosome permitted the mapping of the factor(s) controlling tooth size to within the Yq11 band [15]. The autosomal imbalance in Down's syndrome leads to a marked reduction in permanent tooth size [16, 17], probably due to a reduction in the cellular activity of developing tooth germs. Furthermore, the different responses of teeth, throughout the dentition, to chromosomal aneuploidies probably reflect variations in timing and rate of cell proliferation and differentiation between developing tooth germs [14].

*Occlusion*

Malocclusion due to dental crowding is a heritable defect and results from a discrepancy between tooth size and jaw dimension. The relative etiologic importance of genetics and environment in the determination of dental crowding still remains controversial.

This high frequency of dental crowding in certain populations has been ascribed to racial admixture [18, 19]. However, a large-scale investigation demonstrated that racial crossing produced no increased risk for malocclusion in children [20]. Corruccini and Whitley [21] have shown that increase in malocclusion could not be ascribed to inbreeding. Malocclusion occurs at very low frequency in preindustrial societies, but increases after the introduction of a softer, refined diet [21, 22]. In one or two generations, the prevalence of malocclusion may reach levels equivalent to those found in long-established urban cultures. This increase occurs much too rapidly to be attributed to genetic selection but is consistent with the dietary theory favored by anthropologists [23]. Experimental results obtained with primates fed with artificially softened food favor this theory [24].

The genetic influences on dental crowding mainly concern the controls of the arch dimension [25] and of tooth size [26] which are independent [27, 28]. Although in most cases, the tooth size and arch size are highly correlated [29], suggesting that the two parameters are regulated as parts of an integrated whole, the dental development within individuals is apparently under more rigid control than the growth of the jaws. In pigs, the growth of the jaws during the first year can be greatly reduced by diet manipulation; however, the tooth size and eruption pattern are not as greatly affected. This disproportion, resulting in severe malocclusion [30], may be due to the intrauterine initiation of dental development and the predetermination of tooth size and shape by early primordia which are relatively protected from environmental growth disturbances. Occlusal variations is a polygenic trait, strongly influenced by environmental factors [31].

### *Developmental Aspects*

Odontogenesis begins during the early stages of craniofacial development. Dental primordial cells are provided by neural crest-derived mesenchyme and the oral epithelium of the maxillary and mandibular processes. The developmental sequence integrates cell migration, differential cell division and specific spatial distribution of functional odontoblasts and ameloblasts. All these developmental processes appear to be coordinated by continuous heterotypic and homotypic cell interactions [for reviews, see 32–35]. Cell kinetic-dependent epitheliomesenchymal interactions control the dental lamina formation [36], histomorphogenesis of the enamel organ [37], and odontoblast [38], ameloblast [39] and cementoblast [40] differentiation.

Differential mitotic activities exist during odontogenesis: locally increased mitotic activity was suggested to be involved in the formation of the dental lamina [36]. However, experimental data have shown that the dental lamina resulted from local cell migration and also from change in the orientation of the mitotic spindles [41].

Butler [42] suggested that differential mitotic activity might be involved in specific tooth morphogenesis. In molars, regional variations in the mitotic indices in the inner dental epithelium seem to correlate with cusp formation [43].

Several experimental results led to the hypothesis that a minimal number of cell cycles might be genetically determined in both odontoblasts and ameloblasts before overt differentiation of these cells may occur [44, 45]. The genetic program controlling the number of cell cycles and the coordination of mitotic activities of preodontoblasts and preameloblasts is maintained in vitro, suggesting the existence of an intradental autonomous mechanism of control [7]. Different factors influence the mitotic activity of dental cells, such as cell-matrix interactions [37], transferrin [46] or growth factors [47].

Growth factors are mitogenic polypeptides which also influence cell differentiation and morphogenetic processes during mammalian development [48].

Epidermal growth factor (EGF) accelerates tooth eruption [49–52]. Mouse prepro-EGF mRNA has been localized in the mouth region, in an area normally associated with the incisor, suggesting that EGF may be synthesized by dental cells [53]. However, many effects of EGF on dental cells seem to be indirect: (1) The premature differentiation of odontoblasts and ameloblasts observed in EGF-treated animals might be secondary to mitogenic effects on the proliferative compartment [52]. (2) In vivo, EGF stimulates the mitotic activity of both preodontoblasts and preameloblasts although these cells do not bind iodinated EGF [51]. (3) The proliferation of dental epithelial cells, resulting from EGF treatment, was suggested to be controlled by tissue interactions [54].

Holtzer et al. [55] postulated the existence of quantal division, a mitosis which results in a change in phenotypic expression. Quantal divisions might either be programmed within the cell or be induced by extracellular factors. Thus a growth factor might control cell differentiation simply by governing cell division [56]. However, the activity of growth factors itself is regulated in time and space: all cells are not equivalent in their ability to bind EGF and the binding demonstrates changes during tooth development [54]. An immunological study of the distribution of receptors specific for different growth factors during odontogenesis should provide more precise information about the differential reactivity of dental cells.

The genetic program of odontoblasts and ameloblasts will determine the volume and size of teeth as well as specific phenotypes. The molecular mechanisms of epigenetic controls are not known, but changes in matrix-cell membrane-cytoskeleton interactions are involved [57–62].

Cell-matrix interactions play an important role in odontogenesis by regulating the mitotic activity of the inner dental epithelium [37], by modulating the cytoskeletal organization during odontoblast polarization [63, 64] or by controlling the functional differentiation of ameloblasts [39]. The informative potential of the extracellular matrix probably results from a specific three-dimensional organization of ubiquitous components [65]. According to the model of Ruch and Karcher-Djuricic [57], the influence of the extracellular matrix might be transduced by membrane receptors anchoring cytoskeletal elements.

Differentiation of odontoblasts requires the mesenchymal surface of a specific basement membrane [35, 66]. At the epitheliomesenchymal junction, several matrix constituents change either in their relative amounts or in their distribution during odontoblast differentiation [58, 59, 67]. The redistribution of fibronectin [58, 59] is of particular interest since this molecule interacts with three high-molecular-weight membrane proteins present in the plasma membrane of dental mesenchymal cells [68]. The fibronectin receptor described for several other models [69, 70] has recently been suggested to mediate the attachment of microfilament termini by a specific interaction with talin [71]. Such a transmembrane mechanism might be involved in the redistribution of actin, α-actinin and vinculin that occurs during odontoblast polarization [60]. Changes in the distribution of intermediate filaments [72] and microtubules [73] are also observed during odontoblast polarization.

The influence of the extracellular matrix on gene expression is communicated through the organization of the cytoskeleton [57] and results in regulation of gene expression at all levels: transcription, mRNA processing, translation, posttranslational modifications, secretion and extracellular organization [74]. The cytoskeletal framework, in addition to its role in maintaining cell morphology and in positioning the organelles, might have a major regulatory role in gene expression [75–77].

While the regulatory role of cell-matrix interactions is now well documented, the influence of cell-cell interactions is much more speculative. The presence of true tight junctions between odontoblasts is still an open question [78–80]. Morphological studies of intercellular junctions in odontoblasts has shown the existence of gap junctions [81–83].

Gap junctions in different mature tissues vary in their structure and biochemistry [84]. Physiological studies suggest the presence of more than one functional class of channel in embryonic [85] and differentiated cells [86]. Gap junctions are involved in different functions such as transepithelial permeability, intercellular coupling [87] and cell-to-cell adhesion [88]. Iguchi et al. [83] suggested that the numerous large gap junctions between odontoblasts were involved more in the mediation of information for collagen synthesis and dentine-matrix elaboration than in the transmission of nervous impulses.

Immunological approaches have shown that gap junctions played a crucial role in the transmission of information required for the development of *Xenopus* embryos [89] or for patterning processes in *Hydra* [90]. As gap junctions allow for cell-to-cell communication, they may be important in the establishment and maintenance of spatial patterns during odontogenesis [38].

Cell-to-cell interactions are mediated by cell-adhesion molecules which are involved in morphogenesis [91]. The tissue distribution of cadherins, glycoproteins involved in a $Ca^{2+}$-dependent cell-adhesion mechanism [92], has been studied in developing mouse and chick embryos [93–95]. The spatial and temporal distribution of cadherin subclasses in development has been found to correlate with histogenesis and morphogenesis [91]. Expression of cadherins is dynamic and switches from one subclass to another in relation to morphogenetic events. These different subclasses are not tissue-specific and this implies that specific cell-cell adhesions that occur in embryogenesis are governed by a spatiotemporally regulated expression of a limited number of different classes of cell-adhesion molecules [96]. The genetic control of cadherin expression is still unknown.

Preliminary data demonstrated the presence of L-CAM and N-CAM in tooth germs [97]. However, attempts were not made to block the effects of these cell-adhesion molecules with specific antibodies in order to study their possible role during odontogenesis. This type of experiment has previously demonstrated a role of L-CAM in feather development [98] and the involvement of N-CAM in the migration of motor neurones [99].

The most intriguing questions concern (1) the initiation of odontogenesis and (2) the determinism of dental pattern formation.

*Initiation of Odontogenesis*

Mandibular arches of 9-day-old mouse embryos are capable of forming cartilage and teeth when isolated from the head and explanted, indicating that odontogenetic cells are present in these arches [100, 101]. Odontogenesis is initiated by the mandibular ectoderm, acting on competent but uncommitted neural crest cells [102]. Neural crest cell migration is not a requirement for odontogenic differentiation [102].

The ability to promote tooth formation rapidly decreases when the mandibular epithelium is taken from older embryos [103]. Although there is good evidence for a transfer of information from the epithelial to the mesenchymal compartment, regulated both in time [103] and in space (existence of a diastema), the nature of the signal leading to the commitment of neural crest-derived cells as well as the consequences at the genetic level of this commitment are still unknown.

*Determinism of Dental Pattern Formation*

In order to understand the determinism of dental pattern formation, two speculative models have been formulated: the field model and the clone model.

According to the field model [104], the pattern of a tooth is determined by two independent variables: a series of tooth-forming locations distributed along the jaw, and a continuously graded internal environment that constitutes a morphogenetic field. It was further supposed that the field was divided into three regions, corresponding to the incisors, canines, and cheek teeth, which influenced the tooth buds in different ways. In a later modification of the hypothesis [105], the reactivity of the tooth buds to the field was conceived to be subject to evolutionary changes.

Osborn [106, 107] proposed the clone model, which describes cell growing, differentiating and unfolding a shape without any supracellular control. According to this model, a single clone of homogeneous mesenchymal cells gives rise to four cell clones which, together with the adjacent ectoderm, can develop independently. Three of these clones are dental and will each generate a tooth class (incisors, canines and cheek teeth); the fourth is a basal clone which develops the basal bone of the jaw. Spacing of the teeth results from the generation, by each newly initiated primordium, of an inhibitory influence in adjacent tissue. This model has been tested on the developing lower molar dentition of the mouse embryo, using organotypic cultures or heterotopic grafts of the presumptive site of the first lower molar [108, 109]. The nature of the signal specifying the presumptive site of molar dentition is not known.

The clone theory postulates the existence of determined mesenchymal cells present in the first branchial arches. Many of the homeotic genes of *Drosophila* are known to control cellular determination in a position specific manner along the anteroposterior axis [110, 111]. The importance of these homeobox-containing genes in regulating *Drosophila* development [112, 113] and the conservation and the widespread phylogenetic distribution of the homeobox sequences [114, 115] suggested that these genes could play an important role in vertebrate development [116].

Recently a family of about 15 murine genes [117], some contained in clusters [118, 119], has been recognized, bearing a high degree of homology to *Drosophila* homeoboxes [120, 121]. The activity of some, if not all, of these genes is most likely exerted positively or negatively through direct binding to DNA sequences involved in the control of gene activities.

Mammalian homeobox genes are expressed in early embryo [119, 122] and in some adult stages [122, 123] in a tissue-specific [122] and spatially restricted [124] manner. Hybridization studies indicate a possible role of Hox 1.3 gene product in the inductive processes which are responsible for the formation of lung or stomach [125]. The expression of such homeobox-containing genes might be involved in the determination of the mesenchymal cells specific for the dental clones. However, active homeobox-containing genes have not yet been detected by in situ hybridization within the first branchial arch.

## Conclusion

The different parameters which characterize dentition, such as tooth number, size, shape, class, as well as eruption and occlusion are all genetically programmed. However, the genes which are involved have not been characterized. In mammals, very few mutations are known which affect specific developmental processes and none is specific for odontogenesis. On the other hand, the cloning and characterization of genes affecting development has proved difficult because of the lack of closely linked markers and the large chromosomal regions involved [126]. Furthermore, a single gene may have numerous effects on the phenotype. Pleiotropy reflects the complexity of the action of genes and their subsequent effects through intertwining chemical pathways [127]. For several genes it has been shown that the level of regulation changes during development. Independent members of multigene families also show independent and sometimes contrasting modes of regulation [128]. The specific factors responsible for different modes of regulation for individual genes are not known. Some genes produce regulatory proteins that will specifically affect the action of other genes, thus resulting in a network of cross-reactions and providing a hierarchy of gene functions [127]. Neither these possible direct genetic regulations nor the different epigenetic control mechanisms known to regulate odontogenesis implicate any dental specific gene or gene product except for dentine and enamel. Such a highly complex developmental process as that which, by generating a time-regulated sequential appearance of a defined number of organs with specific shapes and sizes, results in dentition, could be regulated by an epigenetic time- and space-specified coordination of ubiquitous gene products.

## References

1 Butler PM: The odontogeny of mammalian heterodonty. J Biol Buccale 1978;6: 217–227.
2 Brown KS: Evolution and development of the dentition; in Jorgenson RJ (ed): Dentition: Genetic Effects. New York, Liss, 1983, pp 29–66.
3 Burzynski NJ, Escobar VH: Classification and genetics of numeric anomalies of dentition; in Jorgenson RJ (ed): Dentition: Genetic Effects. New York, Liss, 1983, pp 95–106.
4 Alvesalo L, Portin P: The inheritance pattern of missing, peg-shaped and strongly reduced upper lateral incisors. Acta Odontol Scand 1969;27:563–575.
5 Graber LW: Congenital absence of teeth: A review with emphasis on inheritance patterns. J Am Dent Assoc 1978;96:266–275.

6 Shapiro SD, Farrington FM: A potpourri of syndromes with anomalies of dentition; in Jorgenson RJ (ed): Dentition: Genetic Effects. New York, Liss, 1983, pp 129–140.
7 Ahmad N, Ruch JV: Comparison of growth and cell proliferation kinetics during mouse molar odontogenesis in vivo and in vitro. Cell Tissue Kinet 1987;20:319–329.
8 Sakai T: Morphogenesis of molar cusps and tubercles in certain primates; in Kurten B (ed): Teeth: Form, Function and Evolution. New York, Columbia University Press, 1982, pp 307–322.
9 Kirveskari P, Alvesalo L: Dental morphology in Turner's syndrome (45,X females); in Kurten B (ed): Teeth: Form, Function and Evolution. New York, Columbia University Press, 1982, pp 298–303.
10 Alvesalo L: The influence of sex-chromosome genes on tooth size in man. A genetic and quantitative study. Proc Finn Dent Soc 1971;67:3–54.
11 Alvesalo L, Osborne RH, Kari M: The 47,XXY male, Y chromosome, and tooth size. Am J Hum Genet 1975;27:53–61.
12 Kari M, Alvesalo L, Manninen K: Sizes of deciduous teeth in 45,X females. J Dent Res 1980;59:1382–1385.
13 Townsend GC, Alvesalo L: The size of permanent teeth in Klinefelter (47,XXY) syndrome in man. Arch Oral Biol 1985;30:83–84.
14 Townsend GC, Alvesalo L, Jensen B, et al: Patterns of tooth size in chromosomal aneuploidies; in Russell DE, Santoro JP, Sigogneau-Russell D (eds): Teeth Revisited. Proc VIIth Int Symp Dental Morphology. Mem Mus Natl Hist Nat, Series A 1988;53: 25–45.
15 Alvesalo L, de la Chapelle A: Tooth sizes in two males with deletions of the long arm of the Y-chromosome. Ann Hum Genet 1981;45:49–54.
16 Barden HS: Mesiodistal crown size dimensions of permanent and deciduous teeth in Down syndrome. Hum Biol 1980;52:247–253.
17 Townsend GC: Tooth size in children and young adults with trisomy 21 Down syndrome. Arch Oral Biol 1983;28:159–166.
18 Lavelle CLB: Maxillary and mandibular tooth size in different racial groups and in different occlusal categories. Am J Orthod 1972;61:29–37.
19 Horowitz S, Osborne R: The genetic aspects of cranio-facial growth; in Moyers R, Krogman W (eds): Cranio-Facial Growth in Man. Oxford, Pergamon Press, 1971, pp 183–212.
20 Chung CS, Niswander JD, Runck DW, et al: Genetic and epidemiologic studies of oral characteristics in Hawaii's school children. II. Malocclusion. Am J Hum Genet 1971;23:471–495.
21 Corruccini RS, Whitley LD: Occlusal variation in a rural Kentucky community. Am J Orthod 1981;79:250–262.
22 Henrickson PA, Sagne S, Thilander H: Bone, teeth and muscle function. Calcif Tissue Res 1977; 22(suppl):466–467.
23 Beecher RM, Corruccini RS: Effects of dietary consistency on craniofacial and occlusal development in the rat. Angle Orthod 1981;51:61–69.
24 Beecher RM, Corruccini RS, Freeman M: Craniofacial correlates of dietary consistency in a nonhuman primate. J Craniofac Genet Dev Biol 1983;3:193–202.
25 Harris EF, Smith RJ: A study of occlusion and arch width in families. Am J Orthod 1980;78:155–163.
26 Potter RH, Nance WE: A twin study on dental dimension. I. Discordance, asymmetry, and mirror imagery. Am J Phys Anthropol 1976;44:391–396.

27 Keene HJ: Epidemiology study of tooth size variability in caries-free naval recruits. J Dent Res 1971;50:1331–1345.
28 Doris JM, Bernard BW, Kuftinec MM: A biometric study of tooth size and dental crowding. Am J Orthod 1981;79:326–336.
29 McKeown M: Craniofacial variability and its relationship to disharmony of the jaws and teeth. J Anat 1975;119:579–588.
30 Tonge CH, McCance RA: Normal development of jaws and teeth in pigs, and the delay and malocclusion produced by calorie deficiencies. J Anat 1973;115:1–22.
31 King DL: Etiology of malocclusion; in Jorgenson J (ed): Dentition: Genetic Effects. New York, Liss, 1983, pp 83–94.
32 Kollar EJ: Epithelial-mesenchymal interactions: tooth development as a model for instructive induction; in Sawyer RM, Falcon JF (eds): Epithelial-Mesenchymal Interactions in Development. New York, Praeger, 1983, pp 27–56.
33 Ruch JV, Lesot H, Karcher-Djuricic V, et al: Epithelial-mesenchymal interactions in tooth germs: mechanisms of differentiation. J Biol Buccale 1983;11:173–193.
34 Slavkin HC, Snead ML, Zeichner-David M, et al: Concepts of epithelial-mesenchymal interactions during development: tooth and lung organogenesis. J Cell Biochem 1984;26:117–125.
35 Thesleff I, Hurmerinta K: Tissue interactions in tooth development. Differentiation 1981;18:75–88.
36 Gaunt WA: An analysis of the growth of the cheek teeth of the mouse. Acta Anat 1963;54:220–259.
37 Olive M, Ruch JV: Does the basement membrane control the mitotic activity of the inner dental epithelium of the embryonic mouse first lower molar. Dev Biol 1982;93: 301–307.
38 Ruch JV: Odontoblast differentiation and the formation of the odontoblast layer. J Dent Res 1985;64:489–498.
39 Karcher-Djuricic V, Staubli A, Meyer JM, et al: Acellular dental matrices promote functional differentiation of ameloblasts. Differentiation 1985;29:169–175.
40 Andujar MB, Magloire H, Hartmann DJ, et al: Early mouse molar root development: cellular changes and distribution of fibronectin, laminin and type IV collagen. Differentiation 1985;30:111–122.
41 Osman A, Ruch JV: Répartition topographique des mitoses dans les territoires odontogènes de la mâchoire inférieure de l'embryon de souris. J Biol Buccale 1975; 3:117–132.
42 Butler PM: Relative growth within the human first upper permanent molar during prenatal period. Arch Oral Biol 1967;12:983–992.
43 Ruch JV: Determinisms of odontogenesis. Cell Biol Rev 1987;14:1–112.
44 Ruch JV, Karcher-Djuricic V, Gerber R: Quelques aspects du rôle de la prédentine dans la différenciation des adamantoblastes. Arch Anat Microsc Morphol Exp 1972; 61:127–138.
45 Ruch JV, Karcher-Djuricic V, Thiebold J: Cell division and cytodifferentiation of odontoblasts. Differentiation 1976;5:165–169.
46 Partanen AM, Thesleff I: Transferrin and tooth morphogenesis: retention of transferrin by mouse embryonic teeth in organ culture. Differentiation 1987;34:25–31.
47 Brownell AG, Rovero LJ: DNA synthesis of enamel organ epithelium in vitro is enhanced by co-cultivation with non-viable mesenchymal cells. J Dent Res 1980;59: 1075–1080.

48 Carpenter G, Cohen S: Epidermal growth factor. Ann Rev Biochem 1979;48:193–216.
49 Cohen S: Isolation of a mouse submaxillary gland protein accelerating incisor eruption and eyelid opening in the newborn animal. J Biol Chem 1962;237:1555–1562.
50 Hoath SB: Treatment of the neonatal rat with epidermal growth factor: Differences in time and organ response. Pediatr Res 1986;20:468–472.
51 Rihtniemi L, Thesleff I: An autoradiographic study on the effect of epidermal growth factor on cell proliferation in erupting mouse incisors. Arch Oral Biol 1987;32: 859–863.
52 Topham RT, Chiego DJ, Gattone VH, et al: The effect of epidermal growth factor on neonatal incisor differentiation in the mouse. Dev Biol 1987;124:532–543.
53 Rall LB, Scott J, Bell GI, et al: Mouse prepro-epidermal growth factor synthesis by the kidney and other tissues. Nature 1985;313:228–231.
54 Partanen AM, Thesleff I: Localization and quantitation of $^{125}$I-epidermal growth factor binding in mouse embryonic tooth and other embryonic tissues at different developmental stages. Dev Biol 1987;120:186–197.
55 Holtzer H, Rubinstein N, Fellini S, et al: Lineages, quantal cell cycles and the generation of cell diversity. Q Rev Biophys 1975;8:523–557.
56 Mercola M, Stiles CD: Growth factors, superfamilies and mammalian embryogenesis. Development 1988;102:451–460.
57 Ruch JV, Karcher-Djuricic V: On odontogenic tissue interactions; in Slavkin HC, Greulich RC (eds): Extracellular Matrix Influences on Gene Expression. New York, Academic Press, 1975, pp 549–554.
58 Lesot H, Osman M, Ruch JV: Immunofluorescent localization of collagens, fibronectin, and laminin during terminal differentiation of odontoblasts. Dev Biol 1981;82: 371–381.
59 Thesleff I, Barrach HT, Foidart JM, et al: Changes in the distribution of type IV collagen, laminin, proteoglycan and fibronectin during mouse tooth development. Dev Biol 1981;81:182–192.
60 Kubler MD, Lesot H, Ruch JV: Temporo-spatial distribution of matrix and microfilament components during odontoblast and ameloblast differentiation. Roux's Arch Dev Biol 1988;197:212–220.
61 Lesot H, Karcher-Djuricic V, Kubler MD, et al: Membrane-cytoskeleton interactions: inhibition of odontoblast differentiation by a monoclonal antibody directed against a membrane protein. Differentiation 1988;37:62–72.
62 Meyer JM, Lesot H, Karcher-Djuricic V, et al: Membranes basales: Aspects généraux, rôle dans l'odontogenèse. Arch Anat Hist Embryol Norm Exp 1983;66:163–180.
63 Lesot H, Smith AJ, Meyer JM, et al: Cell-matrix interactions: influence of non-collagenous proteins from dentin on cultured dental cells. J Embryol Exp Morphol 1986;96:195–209.
64 Cam Y, Meyer JM, Staubli A, et al: Epithelial-mesenchymal interactions: effects of a dental biomatrix on odontoblasts. Arch Anat Microsc Morphol Exp 1986;75:75–89.
65 Ruch JV, Lesot H, Karcher-Djuricic V, et al: Extracellular matrix-mediated interactions during odontogenesis; in Hinchliff JR, Kemp RB (eds): Matrices and Cell Differentiation. New York, Liss, 1984, pp 103–114.
66 Ruch JV, Lesot H, Karcher-Djuricic V, et al: Facts and hypotheses concerning the control of odontoblast differentiation. Differentiation 1982;21:7–12.

67 Osman M, Ruch JV: $^{3}$H-Glucosamine and $^{3}$H-proline radioautography of embryonic mouse dental basement membrane. J Craniofac Genet Dev Biol 1981;1:95–108.
68 Lesot H, Karcher-Djuricic V, Mark M, et al: Dental cell interaction with extracellular-matrix constituents: type I collagen and fibronectin. Differentiation 1985;29:176–181.
69 Hughes RC, Butters TD, Aplin JD: Cell surface molecules involved in fibronectin-mediated adhesion. A study using specific antisera. Eur J Cell Biol 1981;26:198–207.
70 Akiyama SK, Yamada SS, Yamada KM: Characterization of a 140 kd avian cell surface antigen as a fibronectin binding molecule. J Cell Biol 1986;102:442–444.
71 Horwitz A, Duggan K, Buck C, et al: Interaction of plasma membrane fibronectin receptor with talin – a transmembranous linkage. Nature 1986;320:531–533.
72 Lesot H, Meyer JM, Ruch JV, et al: Immunofluorescent localization of vimentin, prekeratin, and actin during odontoblast and ameloblast differentiation. Differentiation 1982;21:133–137.
73 Ruch JV, Karcher-Djuricic V, Staubli A, et al: Effets de la cytochalasine B et de la colchicine sur les cytodifférentiations dentaires in vitro. Arch Anat Microsc Morphol Exp 1975;64:113–134.
74 Bissell MJ, Barcellos-Hoff MH: The influence of extracellular matrix on gene expression: Is structure the message? J Cell Sci 1987;suppl 8:327–343.
75 Penman S, Fulton A, Capco D, et al: Cytoplasmic and nuclear architecture in cells and tissue: Form, functions and mode of assembly. Cold Spring Harbor Symp Quant Biol 1981;46:1013.
76 Nielsen P, Goelz S, Trachsel H: The role of the cytoskeleton in eukaryotic protein synthesis. Cell Biol Int Rep 1983;7:245–254.
77 Bissell MJ, Hall HG: Form and function in the mammary gland: The role of the extracellular matrix; in Neville MC, Daniel C (eds): The Mammary Gland: Development, Regulation and Function. New York, Plenum Press, 1987.
78 Tanaka T: The origin and localization of dental fluids in developing rat odontoblasts in various stages of their development and maturation. Arch Oral Biol 1980;16:993–1011.
79 Sasaki T, Nakagawa K, Higashi S: Ultrastructure of odontoblasts in kitten tooth germs as revealed by freeze-fracture. Arch Oral Biol 1982;27:897–904.
80 Bishop MA: Evidence for tight junctions between odontoblasts in the rat incisor. Cell Tissue Res 1985;239:137–140.
81 Goldberg M, Escaig F, Septier D: Etude par cryofracture et sur coupes fines des jonctions intercellulaire entre les odontoblastes de l'incisive de rat. Modifications induites par la vinblastine. J Biol Buccale 1981;9:141–153.
82 Calle A: Intercellular junctions between human odontoblasts. A freeze-fracture study after demineralization. Acta Anat 1985;122:138–144.
83 Iguchi Y, Yamamura T, Ichikawa T, et al: Intercellular junctions in odontoblasts of the rat incisor studied with freeze-fracture. Arch Oral Biol 1984;29:487–497.
84 Larsen WJ: Biological implications of gap junctions structure, distribution, and composition: a review. Tissue Cell 1983;15:645–671.
85 Spray DC, White RL, Carvalho AC, et al: Gating of gap junction channels. Biophys J 1984;45:219–230.
86 Margiotta JF, Walcott B: Conductance and dye permeability of a rectifying electrical synapse. Nature 1983;305:52–55.
87 Lawrence TS, Beers WH, Gilula NB: Transmission of hormonal stimulation by cell-to-cell communication. Nature 1978;272:501–506.

88 Staehlin LA: Structure and function of intercellular junctions. Int Rev Cytol 1974; 39:191–283.
89 Warner AE, Guthrie SC, Gilula NB: Antibodies to gap-junctional protein selectively disrupt junctional communication in the early amphibian embryo. Nature 1984;311: 127–131.
90 Fraser SE, Green CR, Bode HR, et al: Selective disruption of gap-junctional communication interferes with a patterning process in *Hydra*. Science 1987;237:49–55.
91 Takeichi M: The cadherins: cell-cell adhesion molecules controlling animal morphogenesis. Development 1988;102:639–655.
92 Takeichi M: Functional correlation between cell adhesive properties and some cell surface proteins. J Cell Biol 1977;75:464–474.
93 Edelman GM, Gallin WJ, Delouvee A, et al: Early epochal maps of two different cell adhesion molecules. Proc Natl Acad Sci USA 1983;80:4384–4388.
94 Nose A, Takeichi M: A novel cadherin adhesion molecule: its expression patterns associated with implantation and organogenesis of mouse embryos. J Cell Biol 1986; 103:2649–2658.
95 Damjanov I, Damjanov A, Damsky CH: Developmentally regulated expression of the cell-cell adhesion glycoprotein cell-CAM 120/80 in peri-implantation mouse embryos and extraembryonic membranes. Dev Biol 1986;116:194–202.
96 Edelman GM: Cell adhesion and morphogenesis: the regulator hypothesis. Proc Natl Acad Sci USA 1984;81:1460–1464.
97 Oliver P, Crossin KL, Slavkin HC: Localization of L-CAM, N-CAM and cytotactin during odontogenic epithelial-mesenchymal interactions. Proc Finn Dent Soc 1987; 83:228.
98 Gallin WJ, Chuong CM, Finkel LH, et al: Antibodies to liver cell adhesion molecule perturb inductive interactions and alter feather pattern and structure. Proc Natl Acad Sci USA 1986;83:8235–8239.
99 Bixby J, Pratt J, Lilien J, et al: Neurite outgrowth on muscle cell surfaces involves extracellular matrix receptors as well as $Ca^{++}$-dependent and independent cell adhesion molecules. Proc Natl Acad Sci USA 1987;84:2555–2559.
100 Lumsden AGS: Determination in early tooth development; in Belcourt AB, Ruch JV (eds): Tooth Morphogenesis and Differentiation. Colloque INSERM, Paris, INSERM, 1985, vol 125, pp 19–27.
101 Gerber R, Karcher-Djuricic V, Ruch JV: Matériel présomptif odontogène et topographie de l'épithélium dentaire dans la mâchoire inférieure de l'embryon de souris. J Biol Buccale 1974;2:249–258.
102 Lumsden AGS: Tooth morphogenesis: Contributions of the cranial neural crest in mammals; in Belcourt AB, Ruch JV (eds): Tooth Morphogenesis and Differentiation. Colloque INSERM, Paris, INSERM, 1985, vol 125, pp 29–40.
103 Mina M, Kollar EJ: The induction of odontogenesis in non-dental mesenchyme combined with early murine mandibular arch epithelium. Arch Oral Biol 1987;32: 123–127.
104 Butler PM: Studies of the mammalian dentition. Differentiation of the postcanine dentition. Proc Zool Soc Lond 1939;109B:1–36.
105 Butler PM: The milk-molars of *Perissodactyla*, with remarks on molar occlusion. Proc Zool Soc Lond 1952;121:777–817.
106 Osborn JW: The evolution of dentitions. Am Sci 1973;61:548–559.

107 Osborn JW: Morphogenetic gradients: fields versus clones; in Butler PM, Joysey KA (eds): Development, Function and Evolution of Teeth. London, Academic Press, 1978, pp 171–201.
108 Lumsden AGS: Pattern formation in the molar dentition of the mouse. J Biol Buccale 1979;7:77–104.
109 Osman M, Karcher-Djuricic V, Ruch JV: Différentiation in vitro du matériel molaire présomptif de l'embryon de souris. C R Acad Sci 1979;289:149–151.
110 Lewis EB: A gene complex controlling segmentation in *Drosophila.* Nature 1978; 276:565–570.
111 Wakimoto BT, Kaufman TC: Analysis of larval segmentation in lethal genotypes associated with the antennapedia gene complex in *Drosophila melanogaster.* Dev Biol 1981;81:51–64.
112 Gehring W: The homeobox: A key to the understanding of development? Cell 1985; 40:3–5.
113 Gehring W: Homeoboxes in the study of development. Science 1987;236:1245–1252.
114 McGinnis W, Garber RL, Wirtz H, et al: A homologous protein-coding sequence in *Drosophila* homeotic genes and its conservation in other metazoans. Cell 1984;37: 403–408.
115 Holland PWH, Hogan BLM: Phylogenetic distribution of antennapedia-like homeoboxes. Nature 1986;321:251–253.
116 Davis CA, Noble-Topham SE, Rossan J, et al: Expression of the homeobox-containing gene En-2 delineates a specific region of the developing mouse brain. Gene Dev 1988;2:361–371.
117 Martin G, Boncinelli E, Duboule D, et al: Nomenclature for homeobox-containing genes. Nature 1987;325:21–22.
118 Colberg-Poley AM, Voss SD, Chowdhury K, et al: Clustered homeoboxes are differentially expressed during murine development. Cell 1985;43:39–45.
119 Hart CP, Awgulewitsch A, Fainsod A, et al: Homeobox gene complex on mouse chromosome II: molecular cloning, expression in embryogenesis and homology to a human homeobox locus. Cell 1985;43:9–18.
120 McGinnis W, Levine MS, Hafen E, et al: A conserved DNA sequence in homeotic genes of the *Drosophila* antennapedia and bithorax complexes. Nature 1984;308: 428–433.
121 Scott MP, Weiner AJ: Structural relationships among genes that control development: Sequence homology between the antennapedia, ultrabithorax and fushi tarazu loci of the *Drosophila.* Proc Natl Acad Sci USA 1984;81:4115–4119.
122 Jackson IJ, Schofield P, Hogan BLM, et al: A mouse homeobox gene is expressed during embryogenesis in adult kidney. Nature 1985;317:745–748.
123 Wolgemuth DJ, Engelmeyer E, Duggal RN, et al: Isolation of a mouse cDNA coding for a developmentally regulated testis-specific transcript containing homeobox homology. EMBO J 1986;5:1229–1235.
124 Awgulewitsch A, Utset MF, Hart CF, et al: Spatial restriction in expression of a mouse homeobox locus within the central nervous system. Nature 1986;320:328–335.
125 Dony C, Gruss P: Specific expression of the Hox 1.3 homeobox gene in murine embryonic structures originating from or induced by the mesoderm. EMBO J 1987; 6:2965–2975.

126 Krumlauf R, Holland PWH, McVey JH, et al: Developmental and spatial patterns of expression of the mouse homeobox gene, Hox 2.1. Development 1987;99:603–617.
127 Bonner JT: The next big problem in developmental biology. Am Zool 1987;27: 715–723.
128 Panduro A, Shalaby F, Shafritz DA: Changing patterns of transcriptional and post-transcriptional control of liver-specific gene expression during rat development. Gene Dev 1987;1:1172–1182.

Hervé Lesot, PhD, Institut de Biologie Médicale, INSERM CJF 88–08,
Faculté de Médecine, 11, rue Humann, F–67085 Strasbourg Cedex (France)

Ferguson DB (ed): Aspects of Oral Molecular Biology.
Front Oral Physiol. Basel, Karger, 1991, vol 8, pp 17–23

# Matrix Genes and Enamel Formation

*J.D. Termine, M.F. Young*

Bone Research Branch, National Institute of Dental Research,
National Institutes of Health, Bethesda, Md., USA

During the formation of enamel, mineral is accumulated as protein is lost in the developing enamel matrix. This change occurs abruptly as the ameloblasts shift from a secretory to a maturational stage of development [1]. During this process, enamel matrix proteins tend to decrease in size and change in chemical character [2–5]. In general terms, the proteins of secretory enamel are hydrophobic and rich in proline, glutamine, leucine and histidine, while the proteins of mature enamel are hydrophilic and rich in acidic amino acids [3–5]. While some matrix protein synthesis does occur during the maturational stage of enamel formation [6], the main function of the maturation ameloblasts is to increase mineral content at the expense of the matrix protein, which becomes degraded and removed from the tissue [2–5].

The proteins contained in erupted enamel are the remnants of this degradative process and are derived from pre-existing acidic matrix constituents in the tissue [7–9]. No trace of the hydrophobic proteins dominant in forming, secretory enamel are found in the posteruptive tooth. The mechanism of this protein degradative process is presumed to be enzymatic and specific to the maturation ameloblast, but the precise enzymes performing this task have not yet been identified.

## *Enamel Matrix Gene Products*

The first clear indication that two separate classes of enamel protein were produced in the developing tooth came when sequential, dissociative extraction procedures were applied to fetal bovine secretory enamel [10, 11]. In this approach, the hydrophobic (proline, glutamine-rich) enamel proteins are almost quantitatively removed from the developing tooth by guanidine HCl extraction, while the hydrophilic (acidic amino acid-rich) enamel proteins are tightly attached to the enamel hydroxyapatite crystals and are only removed on extraction with EDTA. Identical results were obtained later with developing human, pig, sheep, hamster, mouse and rat enamel [3, 4, 12–18]. The hydrophobic, guanidine HCl-soluble proteins are called amelogenins while the crystal-bound, EDTA-soluble proteins are called enamelins [11].

In the very earliest enamel, enamelin synthesis predominates while somewhat later throughout the secretory stage of enamel formation, amelogenin synthesis is dominant [5, 11, 18]. Amelogenin protein is rapidly degraded to smaller fragments and then removed entirely from the maturational enamel, while enamelin degradation is considerably slower with small enamelin fragments persisting in the posteruptive tooth [4, 5, 7, 11, 18]. Enamelin proteins may resist degradation because they are protected by the enamel mineral and are extensively glycosylated [11, 19]. (In general, glycoproteins are more resistant to proteolysis than nonglycosylated proteins.) Further, enamelin proteins are higher in molecular weight (46–72 kd) than their amelogenin counterparts (5–27 kd, with traces of $\approx$40 kd [20] constituents) [11].

Enamelin proteins adsorb strongly to hydroxyapatite and in solution are potent inhibitors of apatite crystal growth [21, 22]. Amelogenin proteins, on the other hand, are less inhibitory [21, 22]. For example, 27 kd amelogenins only appear to inhibit apatite crystal growth via a short acidic segment at their carboxy terminus [32]. In the developing enamel tissue, most of the amelogenin protein is located in spaces between growing enamel mineral crystals, while most of the enamelin protein is adsorbed directly to enamel crystallite surfaces [24]. Thus, enamelin proteins, the first gene products expressed by the early ameloblast [18], may initiate the very first enamel mineral deposits and then regulate growth of the unique, hexagonal rod-like enamel apatite crystals. Amelogenin proteins, the major products of secretory ameloblasts, may contribute to further enamel mineralization by forming a hydrophobic gel in which the mineral crystals are nurtured during the secretory stage of enamel formation. Hydrophobic environments (e.g. with decreased dielectric constants) lower mineral solubility and provide a greater driving force for mineral ion deposition [25]. During the more rapid phase of enamel mineralization that occurs during enamel maturation, the residual enamelin proteins on the crystal surfaces seem to regulate crystal growth, since amelogenin protein is rapidly removed at this stage of tooth formation.

Analysis of developing enamel matrix proteins with monoclonal antibodies directed against either amelogenin [26–28] or enamelin [28, 29] protein indicate that these two classes of enamel matrix proteins are immunologically distinct and probably represent completely different ameloblast gene products. Spectroscopic analysis of isolated amelogenin proteins in solution suggest that this matrix constituent seems to form β-sheet and β-turn structures [30, 31]. However, x-ray diffraction and Fourier transform IR studies both of isolated amelogenin and enamelin protein and of either intact or guanidine HCl-extracted whole enamel tissue showed no regular conformation at all (including β-sheets) for amelogenin protein, but demonstrated clear and distinct β-sheet structure for enamelin protein, both in vitro and in situ [32].

*Primary Sequence and Related Data*

Both isolated protein-derived [33–36] and cDNA-derived [37–39] primary sequence data have been reported for amelogenin proteins (fig. 1). For secreted proteins, all of the 27 kd amelogenins synthesized by ameloblasts in the cow, pig, mouse and man have identical amino terminal sequences [34, 37]. This is quite remarkable when it is realized that each molecular weight form of amelogenin protein (e.g. a single band on SDS gels) has multiple components on charge separation criteria (e.g. on isoelectric focusing). For example, human 27 kd amelogenin is a single band on SDS gel electrophoresis, yet is separable into 7–8 bands on isoelectric focusing [13]. Further, the nine isoelectric focusing bands from the 27 kd bovine amelogenin have identical 33 residue amino termini [34]. The amelogenin proteins are not glycosylated or phosphorylated and thus posttranslational modifications cannot account for their apparent charge heterogeneity [34].

What then is the basis of the heterogeneity of the amelogenin proteins? Cell-free translation of bovine ameloblast mRNA followed by immunoprecipitation with monospecific antiamelogenin antibodies yields 7–9 major products at 27 and 25 kd on two-dimensional gel electrophoresis [26, 38]. Similar SDS gel data (1 dimension only) were obtained for cell-free translated mouse mRNA [40]. On longer exposure of radioautographs, hybrid-selected amelogenin mRNA (i.e. only mRNA that hybridized to cDNA encoding bovine amelogenin) yielded small amounts of 5, 12, and 40 kd bands along with major 25 and 27 kd bands on cell-free translation [38, and unpubl. data].

```
1. MPLPPHPGHPGYINFSYEVLTPLKWYQSMIRHPYPSYGYEPMGGWLHHQI   50
2.                  Q                     Q
3.                                        Q
4.                  Q                     Q

1. IPVVSQQTPQNHALQPHHHIPMVPAQQPVVPQQPMMPVPGQHSMTPTQHH  100
2.           D     P    V                    --------
3.    L   H PS  T      L         A
4           S                 GI        L

1. QPNLPLPAQQPFQPQSIQPQPHQPLQPHQPLQLQPPMQPLQPLQPLQPQP  150
2.          T                     ---              LH
3.    I         QPF-   ----------------AI P SH  M   S
4.          T         G      QPSMHPIQ  L  PP QL PMFS]

1. PVHPIQPLPPQPPLPPIFPMQPLPPMIPDLPLEAWPATDKTKREEVD     172
2.  PQ  VQPI F    Q----     L N   Q  ]
3.  L   M  A        S    S  IL Q  Q                ]
```

*Fig. 1.* Amelogenin amino acid sequence comparison in three species. Row 1: protein sequence deduced from a bovine cDNA [38]; row 2: derived protein sequence from bovine tissue [36]; row 3: deduced protein sequence from a mouse cDNA [37], and row 4: protein sequence determined from porcine tissue [35]. Only amino acids that differ from the bovine cDNA are shown in rows 2–4. Dashes have been inserted for optimal alignment and the brackets delineate the end of each determined sequence.

Northern analysis of bovine and mouse mRNA encoding amelogenin protein again shows multiple molecular species [38, 40, 41]. Low molecular weight amelogenin in tissues is usually thought to arise via proteolysis during enamel maturation. However, direct synthesis of such species cannot be ruled out based on the above mRNA-derived data. For example, a particular 5 kd amelogenin protein, LRAP (leucine-rich amelogenin polypeptide) [34], seems to be composed of the 33 residue 'consensus' amino terminal amelogenin sequence and a 13 residue segment (residues 162–178) found towards the carboxy terminus of the 27 kd amelogenin sequence [38]. It should also be remembered that amelogenin proteins of $\approx$40 kd have been isolated from developing bovine enamel tissue [20, 42]. Since Southern analysis and other data suggest that there is only one amelogenin gene per haploid genome for both the cow [38] and the mouse [40], it is likely that the multiple mRNAs and translated amelogenin proteins described above are formed via alternate gene splicing mechanisms. Interestingly, it has recently been found that antibodies made for a 'universal' 27 residue amino terminal sequence for all amelogenins seem to cross-react with rabbit and mouse enamelin protein [43]. Thus, it may be possible that the enamelin and amelogenin genes, while probably different, are partially homologous, perhaps through exon shuffling.

## *References*


1 Burgess RC, Maclaren C: Proteins in developing bovine enamel; in Stack MV, Fearnhead RV (eds): Tooth Enamel, its Composition, Properties and Fundamental Structure. Bristol, Wright, 1965, pp 74–82.

2 Fukae M, Shimizu M: Studies on the proteins of developing bovine enamel. Arch Oral Biol 1974;19:381–386.

3 Robinson C, Kirkham J, Briggs HD, et al: Enamel proteins: from secretion to maturation. J Dent Res 1982;61:1490–1495.

4 Fincham AG, Belcourt AB, Termine JD: Changing patterns of enamel matrix proteins in the developing bovine tooth. Caries Res 1982;16:64–71.

5 Deutsch D, Shapira L, Alayoff A, et al: Protein and mineral changes during prenatal and postnatal development and mineralization of human deciduous enamel; in Fearnhead RW, Suga S (eds): Tooth Enamel V. Amsterdam, Elsevier, 1984, pp 234–239.

6 Nanci A, Slavkin HC, Smith CE: Immunocytochemical and radio-autographic evidence for secretion and intracellular degradation of enamel proteins by ameloblasts during the maturation stage of amelogenesis in rat incisors. Anat Rec 1987; 217:107–123.

7 Robinson C, Lowe NR, Weatherell JA: Amino acid composition, distribution and origin of 'tuft' protein in human and bovine dental enamel. Arch Oral Biol 1975;20: 29–42.

8 Belcourt A, Gillmeth S: EDTA soluble protein of human mature enamel. Calcif Tissue Int 1979;28:227–231.

9 Belcourt AB, Fincham AG, Termine JD: EDTA insoluble protein of adult human enamel. Caries Res 1982;16:72–76.

10 Termine JD, Torchia DA, Conn KM: Enamel matrix: structural proteins. J Dent Res 1979;58B:773–778.

11 Termine JD, Belcourt AB, Christner PJ, et al: Properties of dissociatively extracted fetal tooth matrix proteins. I. Principal molecular species in developing bovine enamel. J Biol Chem 1980;255:9760–9768.

12 Lyaruu DM, Belcourt A, Fincham AG, et al: Neonatal hamster molar tooth development: extraction and characterization of amelogenins, enamelins and soluble dentin proteins. Calcif Tissue Int 1982;34:86–96.

13 Fincham AG, Belcourt AB, Termine JD: Molecular composition of the protein matrix of developing human dental enamel. J Dent Res 1983;62:11–15.

14 Zeichner-David M, MacDougall M, Vides J, et al: Immunochemical and biochemical studies of human enamel protein during neonatal development. J Dent Res 1987; 66:50–56.

15 Fincham AG, Belcourt AB, Lyaruu DM, et al: Comparative protein biochemistry of developing enamel from five mammalian species. Calcif Tissue Int 1982;34:182–189.
16 Limeback H: Isolation and characterization of pig enamelins. Biochem J 1987; 243:385–390.
17 Fukal M, Tanaki T: Nonamelogenin components of porcine enamel in the protein fraction free from the enamel crystals. Calcif Tissue Int 1987;40:286–293.
18 Slavkin HC: Gene regulation in the development of oral tissues. J Dent Res 1988; 67:1142–1149.
19 Menanteau J, Meflah K, Strecker G: The carbohydrate moiety of mineral bound proteins from fetal enamel: a basis for enamelins heterogeneity. Calcif Tissue Int 1988;42:196–200.
20 Belcourt AB, Fincham AG, Termine JD: Bovine high molecular weight amelogenin proteins. Calcif Tissue Int 1983;35:111–114.
21 Doi Y, Eanes ED, Shimokawa H, et al: Inhibition of seeded growth of enamel apatite crystals by amelogenin and enamelin proteins in vitro. J Dent Res 1984;63:98–105.
22 Cribbin NJ, Shore RC, Robinson C, et al: Control of crystal growth during maturation of rat incisor enamel. J Dent Res 1988;67:642.
23 Aoba T, Fukae M, Tanake T, et al: Selective adsorption of porcine amelogenins onto hydroxyapatite and their inhibitory activity on hydroxyapatite growth in supersaturated solutions. Calcif Tissue Int 1987;41:281–289.
24 Hayashi Y, Bianco P, Shimokawa H, et al: Organic-inorganic relationships and immunohistochemical localization of amelogenins and enamelins in developing enamel. Basic Appl Histochem 1986;30:291–299.
25 Termine JD, Posner AS: Calcium phosphate formation in vitro. I. Factors affecting initial phase separation. Arch Biochem Biophys 1970;140:307–317.
26 Shimokawa H, Wassmer P, Sobel ME, et al: Characterization of cell-free translation products of mRNA from bovine ameloblasts by monoclonal and polyclonal antibodies; in Fearnhead RW, Suga S (eds): Tooth Enamel IV. Amsterdam, Elsevier, 1984, pp 161–166.
27 Christner PJ, Lally ET, Ads AH, et al: Production of a monoclonal antibody to bovine tooth enamel proteins. Arch Oral Biol 1983;18:773–779.
28 Deutsch D, Palmon A, Catalano-Sherman J, et al: Production of monoclonal antibodies against enamelin and amelogenin proteins of developing enamel matrix. Adv Dent Res 1987;1:282–288.
29 Rosenbloom J, Lally E, Dixon M, et al: Production of a monoclonal antibody to enamelins which does not cross-react with amelogenins. Calcif Tissue Int 1986;39: 412–415.
30 Renugopalakrishnan V, Strawich ES, Horowitz PM, et al: Studies of the secondary structure of amelogenin from bovine tooth enamel. Biochemistry 1986;25:4879–4887.
31 Zhenz S, Tu AT, Renugopalakrishnan V, et al: A mixed β-turn and β-sheet structure for tooth enamel amelogenin: Raman spectroscopic evidence. Biophysics 1987;26: 1809–1813.
32 Jodaikin A, Weiner S, Perl-Treves D, et al: Developing enamel matrix proteins: a conformation study of enamelins and amelogenins. Int J Biol Macromolecules 1987;9:166–168.
33 Fincham AG, Belcourt AB, Termine JD, et al: Dental enamel matrix: sequences of two amelogenin polypeptides. Biosci Rep 1981;1:771–778.

34 Fincham AG, Belcourt AB, Termine JD, et al: Amelogenins: sequence homologies in enamel matrix proteins from three mammalian species. Biochem J 1983;211:149–154.
35 Fukae M, Shimazu M: Amino acid sequence of the main component of porcine enamel proteins. Jpn J Oral Biol 1983;25(suppl):29.
36 Takagi T, Suzuki M, Baba T, et al: Complete amino acid sequence of amelogenin in developing bovine enamel. Biochem Biophys Res Commun 1984;121:592–597.
37 Snead ML, Lau EC, Zeichner-David M, et al: DNA sequence for cloned cDNA for murine amelogenin reveal the amino acid sequence for enamel-specific protein. Biochem Biophys Res Commun 1985;129:812–818.
38 Shimokawa H, Sobel ME, Sasaki M, et al: Heterogeneity of amelogenin mRNA in the bovine tooth germ. J Biol Chem 1987;262:4042–4047.
39 Shimokawa H, Ogata Y, Sasaki S, et al: Molecular cloning of bovine amelogenin cDNA. Adv Dent Res 1987;1:293–297.
40 Snead ML, Lau EC: Examining the possible molecular origins for enamel protein complexity. Adv Dent Res 1987;1:298–305.
41 Young MF, Shimokawa HS, Sobel ME, et al: A characterization of amelogenin messenger RNA in the bovine tooth germ. Adv Dent Res 1987;1:289–292.
42 Christner PJ, Lally ET, Miller RD, et al: Monoclonal antibodies to different epitopes in amelogenins from fetal bovine teeth recognize high molecular weight components. Arch Oral Biol 1985;30:849–854.
43 Zeichner-David M, Vides J, MacDougall, et al: Biosynthesis and characterization of rabbit tooth enamel extracellular matrix proteins. Biochem J 1988;251:631–641.

J.D. Termine, MD, Bone Research Branch, National Institute of Dental Research, National Institutes of Health, Bethesda, MD 20982 (USA)

Ferguson DB (ed): Aspects of Oral Molecular Biology.
Front Oral Physiol. Basel, Karger, 1991, vol 8, pp 24–41

# Phylogenetic Expression of Enamel Proteins

## Immunologic Detection and Localization of Amelogenins and Enamelins in Vertebrate Teeth[1]

*Richard Herold, Joel Rosenbloom*

Anatomy and Histology Department and Research Center in Oral Biology, School of Dental Medicine, University of Pennsylvania, Philadelphia, Pa., USA

Current understanding of the genetic control of enamel formation is rather rudimentary partly because information is just being developed on identification, timing of expression, the differential functions of the various enamel proteins, and how these relate to the complex process of amelogenesis. In order to investigate the genetic control of enamel formation it is necessary to be able to determine when and where the multiple enamel protein genes are expressed during amelogenesis. Such analyses require probes which are capable of identifying either specific mRNAs or the proteins themselves. Complementary DNA probes for mouse [1] and bovine [2] amelogenin have been constructed and characterized and these are described in detail elsewhere in this volume. A promising avenue for identifying and localizing the individual enamel proteins has been the development and use of monoclonal antibodies which react with single antigenic determinants. The main thrust of this chapter will be to review the use of such immunoprobes to identify and localize enamel proteins in developing tooth enamel in a wide variety of vertebrates, and to relate these results to possible genetic control of enamel formation.

Enamel proteins in developing mammalian enamel were initially studied by extracting proteins from the enamel layer with various solutions. Depending upon the extraction method and the stage of enamel development, mixures of proteins were obtained which contained from 4 to 20 components, which varied in molecular weight from 2,000 to 70,000 [3–10]. These

[1] Supported by NIH Grants DE08063 and DE08239.

data were interpreted to mean that a relatively few enamel proteins of relatively high molecular weight were synthesized initially by the ameloblasts and secreted into the extracellular matrix. Subsequently, during the mineralization and maturation processes, these proteins were broken down into smaller molecules. These smaller molecular weight proteins were then removed from the matrix with the result that the mature highly calcified enamel matrix contained only a small amount of protein.

In a very important paper in 1980, Termine et al. [11] demonstrated that enamel proteins could be operationally divided into two groups on the basis of their solubility in guanidine hydrochloride (GuHCl) with or without EDTA. These two groups differ in amino acid composition and molecular weights. The majority of the protein in the group soluble in GuHCl alone, designated amelogenins, has molecular weights in the range of 10,000–30,000, although small amounts of higher molecular weight components (40,000) have been identified [12, 13]. In contrast, the majority of the protein in the fraction soluble in GuHCl plus EDTA, designated enamelins, has molecular weights greater than 40,000. During enamel maturation, it is apparently the amelogenins which are largely removed, while the enamelins which are tightly bound to the hydroxyapatite crystals are largely retained [14, 15]. Presumably, these two classes of proteins are involved in different ways in the formation of the mature enamel matrix.

In order to investigate the role of the subclasses of enamel proteins, we have developed monoclonal antibodies which can distinguish the two classes [12, 16], and have used them to study the distribution of amelogenins and enamelins in the developing enamel layer of a variety of vertebrates extending from fish to man.

## *Preparation and Testing of Immunoprobes for Enamel Proteins*

### *Generation of Polyclonal Antibodies to Bovine Enamel Proteins*

In our initial studies [17] we generated polyclonal antibodies in rabbits to enamel proteins extracted with 0.5 *M* EDTA from molar teeeth of 6-month-old fetal calves. The extract contained approximately 95% of the total protein found in the tissue and included both amelogenins and enamelins. Subsequent analyses by Western blotting techniques using polyclonal antibodies generated against this mixture of proteins, demonstrated that the antibody reacted with both amelogenins and enamelins, but not with other antigens such as collagen or bovine serum albumin. The antibodies were also tested by immunofluorescent techniques on sections prepared from unerupted deciduous molar teeth from 4-month-old fetal calves. At this age, the teeth are in the bell stage of development and all stages of extracellular matrix formation can be seen in longitudinal sections. Fluorescence was detected in the cytoplasm of the preameloblasts and to a limited extent in the stratum intermedium cells before the deposition of any enamel matrix. In the area of the teeth where enamel deposition was occurring, immunofluorescent staining was distributed throughout the newly secreted matrix. In some areas of the teeth, the apical ends of the ameloblasts showed fluorescence in Tomes' processes. In the region where enamel matrix had obtained a thickness of 200 µm, a 20-µm thick layer near the surface was brightly and uniformly fluorescent, and a more faintly fluorescent 30-µm layer was found near the dentinoenamel junction. Between the two fluorescent layers, a thick non-fluorescent layer of enamel matrix was present. No fluorescence was evident in any other tissues of the tooth germ nor in various embryonic tissues containing epithelial tissue including liver, salivary glands and pancreas. In addition, no fluorescence was evident in actively secreting odontoblasts which contain large amounts of procollagen, in tongue epithelium which contains large amounts of keratin or in the pulp and stellate reticulum of the tooth bud which contain large amounts of proteoglycans. These results supported the relatively specific nature of our antibody preparation. When normal rabbit serum was used in place of the specific antibody, no immunofluorescence was observed.

In order to determine whether our polyclonal antibody to bovine enamel protein would cross-react with enamel proteins of other species, developing tooth buds from a variety of mammalian and reptilian species as well as the tooth-covering tissue, enameloid, of lower vertebrates were examined by histologic immunofluorescent techniques. We found that cross-reactive proteins were present in the enameloid of sharks, fish and amphibians as well as in the enamel of reptiles and all mammals tested including mouse, pig and man [18]. These results suggested that at least part of the structure of enamel proteins has been strongly conserved throughout evolution. The presence of enamel proteins in the enameloid of shark dermal denticles suggests that the earliest tooth-like structure in vertebrates, the odontodes of the dermal armor, probably were also covered by enameloid, which contains enamel-like proteins. Furthermore, the presence of enamel proteins in both enamel and enameloid tissues suggest that they play a central role in the development of these extremely hard tissues, possibly through involvement in the mineralization process.

*Production of Monoclonal Antibodies to Enamel Proteins*

Because the polyclonal antisera were developed against impure mixtures of enamel proteins, it was impossible to use such antisera in the precise classification of enamel proteins or in determining which enamel proteins are present during the secretory or maturation phases of amelogenesis. In order to use antienamel protein antibodies to identify and classify enamel protein precursors and their degradation products, it is desirable that these probes be directed against a single antigenic site on the enamel protein. To this end, we have developed monoclonal antibodies which react specifically with either the amelogenin or the enamelin protein fractions. We used essentially the techniques described originally by Koehler and Milstein. Amelogenin or enamelin proteins isolated from fetal bovine enamel were linked to keyhole limpet hemocyanin (KLH) by cross-linking with glutaraldehyde. Six-week-old female Balb/C mice were injected intraperitoneally with 100 μg of KLH-enamel proteins (a mixture of amelogenins and enamelins) biweekly. After three injections, the mice were bled from the retro-orbital plexus and the sera tested for the presence of antibodies in an enzyme-linked immunosorbent assay (ELISA). Three days prior to spleen cell hybridization, the mice were given an intravenous injection of 400 μg of the antigen and those mice whose sera were positive in the ELISA at a dilution of 1:5,000 were used in the fusion.

Single-cell suspensions were obtained by homogenization of the spleens with a loose fitting tissue homogenizer in serum-free RPMI-1640 medium. These cells were fused to the myeloma cell line Sp2/0-Ag[14] using polyethylene glycol (ratio of 1:5, spleen:myeloma). The fusion mixture was centrifuged and the cells resuspended in Dulbecco's medium containing 10% NCTC medium, 20% horse serum, 0.15 mg/ml oxaloacetate, 0.05 mg/ml pyruvate, 0.2 μg/ml bovine insulin and gentamycin (10 μg/ml). The cell mixtures were plated into 96-well tissue culture plates and the following day 0.5 ml of media containing $4 \times 10^{-7}$ *M* aminopterin was added. Cultures were fed every 3 days and clonal growth was detected after 7 days. Clones judged positive by ELISA were subcloned by plating $10^2$–$10^3$ cells in Sea Kem Agarose in a 100-mm Petri plate. After 2 weeks, clones were picked from the agarose with a Pasteur pipette, grown up and assayed for antienamel protein antibody.

*Antiamelogenin Monoclonal Antibodies*

To date, six monoclonals to amelogenins have been purified and characterized in detail [12, 13]. When the amelogenins in the GuHCl extract were fractionated by electrophoresis on sodium dodecyl sulfate (SDS)-urea poly-

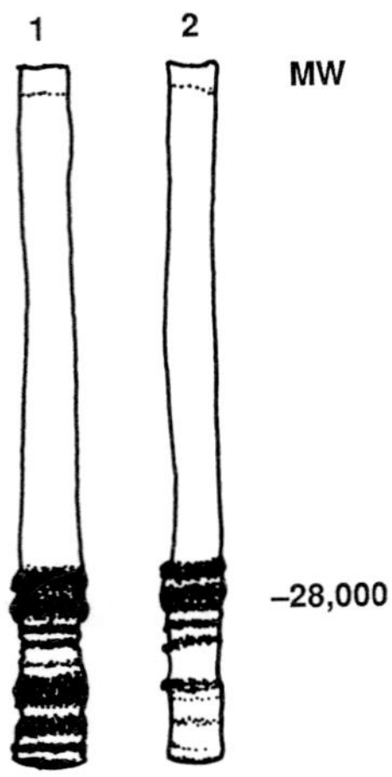


1

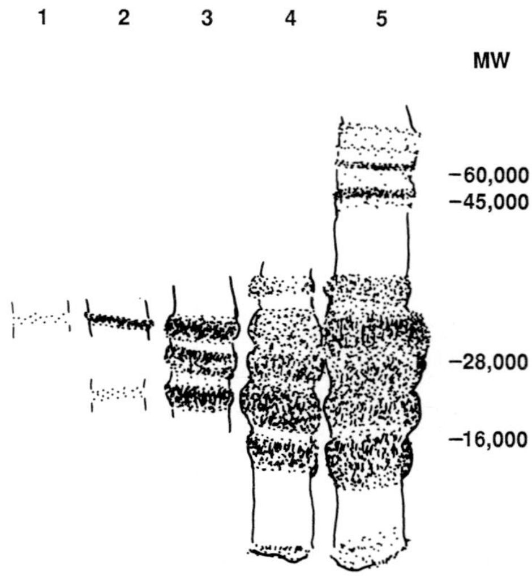


2

*Fig. 1.* SDS Laemmli slab gel of fetal bovine amelogenins. Lane 1, 200 μg. The gel was stained with Coomassie blue. A representative molecular weight is shown.

*Fig. 2.* Immunoblot of fetal bovine amelogenins electrophoresed on a Laemmli SDS slab gel and transferred onto nitrocellulose. Amelogenins shown here were localized by antibody 316 and peroxidase-conjugated rabbit antimouse antibody. Lanes 1–5 contained the following amounts of amelogenin protein: 0.04, 0.2, 1.0, 5.0 and 25 μg, respectively

acryl-amide gels, four major components in the range 16,000–30,000 daltons were visualized by Coomassie blue staining in addition to several other minor components (fig. 1). After transfer to nitrocellulose, all the major components reacted with all six antibodies (fig. 2). In addition, when large amounts, 25–100 μg, of amelogenins were electrophoresed, two higher molecular weight components not seen with the Coomassie blue, were visualized. The molecular weights of these two components were approximately 45,000 and 60,000 (fig. 2, lane 5).

The high molecular weight species localized by the monoclonal antibodies had molecular weights similar to that reported for enamelins. It was, therefore, important to determine if these protein bands were enamelins or if they represented larger amelogenin molecular species. Enamelins were electrophoresed using the Laemmli gel system (fig. 3). Two lanes were stained with silver stain and three lanes were blotted to nitrocellulose and incubated

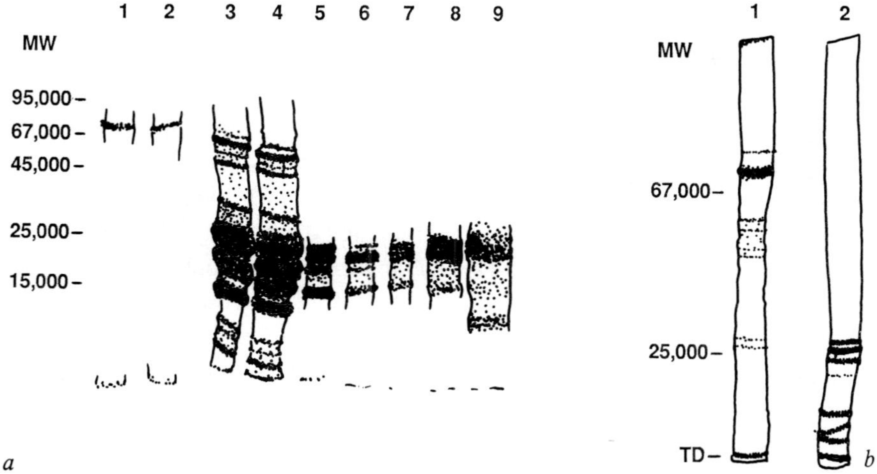


*Fig. 3.* *a* SDS Laemmli slab gel and immunoblot. Lanes 1 and 2 are a Laemmli SDS slab gel of fetal bovine enamelins stained with silver stain. Lanes 3–9 are an immunoblot of the same gel. In lanes 3–6: 25, 6.25, 1.5 and 0.4 μg of fetal bovine amelogenins were electrophoresed, respectively. In lanes 7–9: 6.25, 25 and 100 μg of enamelins were elctrophoresed. The protein bands recognized by antibody 316 were visualized by incubation with peroxidase-conjugated rabbit antibody to mouse antibody. MW = Molecular weight. *b* Transfer of enamelins and amelogenins to nitrocellulose after SDS electrophoresis. Proteins were subject to electrophoresis and transferred to nitrocellulose as described in 'Materials and Methods'. After transfer, they were stained with amido Schwartz dye. Lane 1: enamelins; lane 2: amelogenins.

with monoclonal antibodies numbered 10, 121 or 316. In figure 3, antibody localization with 316 is shown; no enamelin proteins were immunolocalized in the molecular weight range where enamelins were stained with silver stain (lanes 1 and 2). The protein localized in lanes 7–9 is in the molecular weight range for the major components of the amelogenins and is presumably the localization of low-level amelogenin contamination in our enamelin preparation. The other antibodies tested (121 and 10) gave identical results (not shown). Because it could be argued that the enamelins did not transfer to nitrocellulose, preparations of amelogenins and enamelins were blotted on nitrocellulose and then localized on the blot with amido Schwartz dye (fig. 3). It can readily be seen that the enamelins transferred well, but were not recognized by these monoclonal antibodies.

*Classification of Antiamelogenin Monoclonal Antibodies into Groups Recognizing Three Different Epitopes*

Since six monoclonal antibodies were obtained, it was of interest to determine whether they recognized different or identical antigenic sites. A competition assay was constructed in which each antibody was tested against the other five. Each of the monoclonal antibodies was allowed to react with bound amelogenins in separate microtiter wells and then the wells were incubated with one of the antibodies conjugated directly to alkaline phosphatase to determine if prior binding of any other antibody prevented the binding of the conjugated antibody. The presumption was that if the first nonconjugated monoclonal antibody was bound to the same or an overlapping antigenic site as that recognized by the conjugated antibody, then the conjugated antibody could not bind to the amelogenin molecules.

The results of these studies can be summarized as follows: antibodies 98 and 181 compete with all the antibodies; antibodies 10, 121, and 316 compete with all the antibodies except 185, antibody 185 competes only with 98 and 181. Thus, the antibodies fall into three groups and a model can be developed to explain the results. Protein epitopes may be divided into two classes: those that lose their antigenicity upon denaturation of the protein and those which retain their antigenicity. Although different terms have been used to distinguish the two classes, such as conformational versus sequential [19], discontinuous versus continuous [20], and assembled topographic versus segmental [21], the retention of antigenic activity under denaturing conditions implies that the epitope is composed of a continuous sequence of amino acids and the retention of a particular conformation is not critical for reactivity. As the amelogenin protein used as antigen was extracted with high concentrations of GuHCl and as the amelogenin components exhibited reactivity with the antibodies after electrophoresis under denaturing conditions, it is likely that the epitopes are continuous in type. With this assumption, a simple model (fig. 4) can be developed to explain the competition

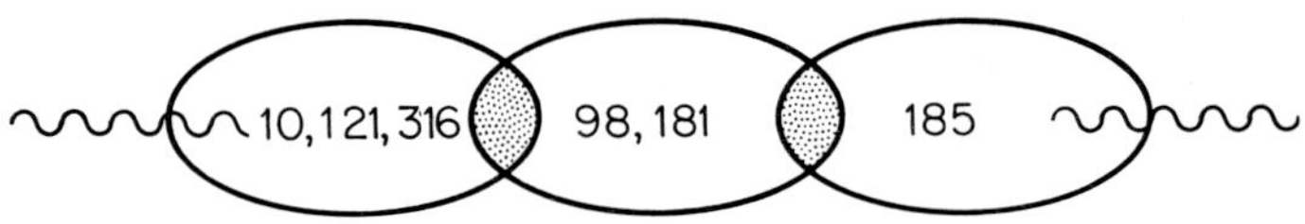


*Fig. 4.* Model of the relationship of amelogenin epitopes defined by monoclonal antibodies. The epitopes are depicted by ovals, with the numbers of the reactive monoclonal antibodies shown within the ovals. This model is based upon the assumption that the epitopes are of the continuous type.

experiments. The group of antibodies composed of 98 and 181 compete with both other groups that do not compete with each other. This implies either that the epitope reactive with 98 and 181 shares sequences with both the other two epitopes, or that the binding of 98 and 181 sterically hinders the binding of the other antibodies. In this continuous model, the epitope to 98 and 181 must be between the other two epitopes. Although the model is the simplest explanation, the data do not preclude some variation in the antigenic determinants recognized by each antibody group. For example, if 10, 121 and 316 do not recognize an identical antigenic determinant, their respective determinants must heavily overlap and differ by a relatively few amino acids.

*Antienamelin Monoclonal Antibodies*

Only one clone reacting with enamelin protein has been characterized in detail so far [16]. In order to characterize the specificity of this antienamelin antibody, it was tested by immunoblotting against enamelin and amelogenin fractions. The results shown in figure 5 clearly demonstrate reactivity with

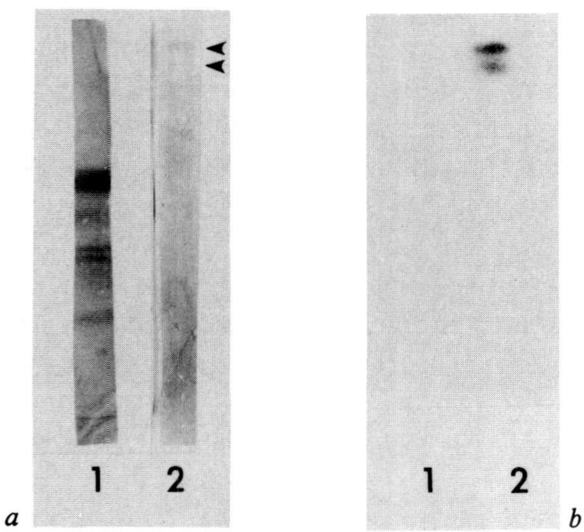


*Fig. 5.* Immunoblotting of 10 μg amelogenin and 10 μg of enamelin proteins. *a* Amelogenin (lane 1) and enamelin (lane 2) proteins were subject to SDS gel electrophoresis and then transferred electrophoretically to nitrocellulose. The nitrocellulose filters were then stained with Coomassie blue. Arrowheads mark the position of immunologically stained bands in *b*. *b* After electrophoretic transfer, the fractionated amelogenins (lane 1) and enamelins (lane 2) were tested for immunologic reactivity by incubation with monoclonal antibody to enamelin protein, followed by incubation with $^{125}I$ protein A and autoradiography.

two high molecular weight components of approximately 68,000 and 72,000 in the enamelin preparation (fig. 5b, lane 2). It should be noted that although two high molecular weight bands were visualized using the antibody, only one band was well visualized by staining for protein with Coomassie blue (fig. 5a, lane 2). At present we have no conclusive explanation for this observation.

It is possible that the clearly stained band is heterogeneous and contains more than one component, only one of which reacts well with the Coomassie blue. Alternatively, the reactive antigenic determinant in the stained protein may be partially blocked by glycosylation or other mechanisms which lessen the reactivity. Further studies are necessary to resolve this issue. When the amelogenin fraction was tested for reactivity, none of the components present in the preparation reacted with the antibody (fig. 5b, lane 1). This lack of reactivity with the amelogenins was not due to absence of transfer to the nitrocellulose since appropriate amelogenin protein bands could easily be visualized on the nitrocellulose by Coomassie blue staining (fig. 5a, lane 1).

*Amelogenins and Enamelins Are Immunologically Distinct Proteins*

This work with the monoclonals suggests that the proteins operationally defined as enamelins and amelogenins do not share antigenic determinants. However, before this conclusion can be reached with reasonable certainty it is necessary to characterize these antigenic determinants. Since the proteins we are studying have been extracted with high concentrations of the chaotropic agent GuHCl and since the proteins exhibited immunoreactivity after electrophoresis under denaturing condition, it is likely that each epitope is composed of a continuous sequence of amino acids rather than being assembled from amino acids located in different parts of the proteins. This suggests that it may be possible to define the amino acid sequences in the antigenic determinants.

## *Application of Immunoprobes to Determine Relation of Various Enamel Proteins to Enamel Formation*

### *Expression of Genes for Enamel Proteins*

Initial studies of the immunohistochemical distribution of enamel proteins were carried out using polyclonal antibodies prepared against extracted enamel proteins containing both amelogenins and enamelins [17, 18, 22, 23]. These studies clearly showed that enamel proteins are present in the ameloblast cells as well as in the enamel matrix during amelogenesis, demonstrating that the enamel proteins are synthesized in the ameloblasts and transported to the forming extracellular enamel matrix. Similar processes have been shown to occur in a large number of mammalian species. The enamel proteins have been shown to be very similar in all these species by amino acid analysis and by cross-reaction of antisera prepared against enamel protein of one species against the enamel proteins of all other species tested [18, 24].

Antiamelogenin and antienamelin monoclonal antibody probes described above were used to show expression of the proteins during amelogenesis by immunochemical, immunohistochemical and immunocytochemical techniques [12, 16, 25]. Both amelogenins and enamelins were detected in the cisternae of the rough endoplasmic reticulum, Golgi saccules and secretory granules of ameloblasts, suggesting that they are merocrine secretions. Within the cells, both amelogenins and enamelins appear to be localized together within the same organelles. After release from the cells into the extracellular enamel matrix, the proteins are found in stippled material and in the calcified matrix [26–33]. It has been suggested that the amelogenin is localized in the organic matrix between crystals and the enamelin is located within the crystals. Unequivocal demonstration of this localization has not been accomplished [34].

The sequential expression of enamel proteins has been described during mouse molar amelogenesis using two-dimensional gel electrophoresis [35, 36]. The earliest expressed enamel protein was detected at fetal age 18 days and had a molecular weight of 46 kd. One day later an additional 72-kd protein appeared, and finally a third protein of 26 kd was detected. Since some enamelin components are reported to be around 72 kd and some amelogenins around 26 kd, these data suggest the possibility that enamelin is synthesized and/or secreted before amelogenin. The 46-kd protein remains to be identified and characterized.

*Expression of Enamel Proteins in Nonmammalian Vertebrate Tooth-Covering Tissue*

The distribution of enamel proteins in nonmammalian vertebrates was initially investigated using polyclonal antibodies made against bovine enamel proteins [37, 38]. This study demonstrated that the tooth-covering tissues (enamel or enameloid) of all nonmammalian vertebrate teeth and dermal denticles contained enamel proteins of one or all the multiple types and that these proteins were produced by the inner enamel epithelial cells (ameloblasts) and secreted into the extracellular matrix of the tooth-covering tissues. These results have been corroborated and extended to other nonmammalian species by Slavkin and co-workers [39–42].

Nonmammalian vertebrate teeth were re-examined using our monoclonal antibodies to enamelin and amelogenin. Surprisingly these data indicated that although all nonmammalian as well as mammalian tooth-covering tissues contained enamelin, only teeth of adult amphibians, reptiles and mammals contained amelogenins [43, 44]. Larval amphibian teeth did not contain amelogenin. Two-dimensional electrophoresis of extracts of shark teeth also indicated the lack of amelogenin proteins [42]. The lack of amelogenin protein in any of the bony and cartilaginous fish examined in these studies strongly suggests that the gene for this protein is missing from the genome of these animals.

The distribution of enamel proteins in amphibian teeth is particularly interesting from the viewpoint of genetic control of enamel formation. Most amphibians have a free-living larval aquatic stage which metamorphoses into a terrestrial adult stage [45–48]. The metamorphosis from larva to adult is dramatic, usually involving the loss of prominent external gills. Many other profound changes occur with metamorphosis which are less obvious, such as the change from ammotelic to urotelic nitrogen metabolism [49]. Extensive changes also occur in the teeth of the forms which have true teeth in both the larval and adult stages [46]. Detailed data on the distribution of amelogenins and enamelins in the tooth-covering tissues of representative amphibians including those having true teeth only in the adult stage (anurans), those with true teeth in both the larval and adult stages (urodeles) and neotenous forms of urodeles [47] which become sexually mature while retaining larval characteristics are presented in figures 6–9 and table 1.

In the anurans (frogs), the larvae do not have true teeth, only horny, keratinized mouth projections, while the adults have numerous calcified bicuspid teeth with hard surface-covering tissue (enamel) and dentin cores [45]. In the adult bullfrog teeth, the surface-covering tissue contains both enamelin and amelogenin (table 1; fig. 6a). In the urodeles (salamanders), true teeth are present in both the larval and adult forms. However, the larval teeth are monocuspid while the adult teeth are bicuspid, and both are covered by a highly calcified surface-covering tissue (larvae, enameloid; adult, enamel) [45]. The change in the shape of tooth formed by the epithelial enamel organ, mono- or bicuspid, occurs during metamorphosis. The enamel protein content of the tooth surface-covering tissues also changes from larval to adult teeth. Both contain enamelin but only the adult teeth contain amelogenin (table 1; fig. 8a, c, 9a–c). Before metamorphosis, the inner enam-

*Table 1.* Distribution of enamel proteins, amelogenin and enamelin, in tooth-covering tissue of representative amphibians

| Amphibians | Stage | Immunogold-silver staining[1] | | |
|---|---|---|---|---|
| | | enamelin[2] | amelogenin[3] | control |
| Bullfrog *Rana catesbiana,* Anura | adult | + | + | 0 |
| Mudpuppy *Necturus maculosus,* Urodele | neotenous form | + | 0 | 0 |
| Spotted salamander | larval | + | 0 | 0 |
| *Ambystoma maculatum,* Urodele | adult | + | + | 0 |

[1] DeMey et al. [55].
[2] Rosenbloom et al. [16].
[3] Christner et al. [12].

el cells only produce enamelin for transport into the extracellular matrix. After metamorphosis, the cells produce both enamelin and amelogenin. The genome of the salamander *(Ambystoma)* must contain the genes for both enamelin and amelogenin but expression of the genes is dependent on the stage of development. Certain salamanders, for example the mudpuppy *Necturus*, show neoteny in which they do not undergo metamorphosis but retain larval characteristics throughout life except for the gonads which become fully functional. Mudpuppy teeth retain the typical monocuspid larval throughout life. These teeth contain only enamelin protein in the tooth-covering tissue (table 1; fig. 7a–c).

The effects of amphibian metamorphosis on differentiation and development of the various organ systems has been the object of investigations for many years. Thyroid hormones have been shown to play a major role in this process [49]. Early work involving the implantation of thyroid hormone pellets to induce precocious local metamorphic change demonstrated that the action of thyroid hormones is tissue specific or that the metamorphic response is an inherent property of a particular tissue. Investigations into the mechanism of thyroid hormone induction of metamorphosis have implicated the mitochondria, plasma membrane, protein synthetic machinery and nucleus as direct sites of action [50]. More recently however, the identification of a chromatin-bound thyroid hormone receptor suggests that the main site of thyroid hormone action is on the nucleus [50]. This nuclear receptor is

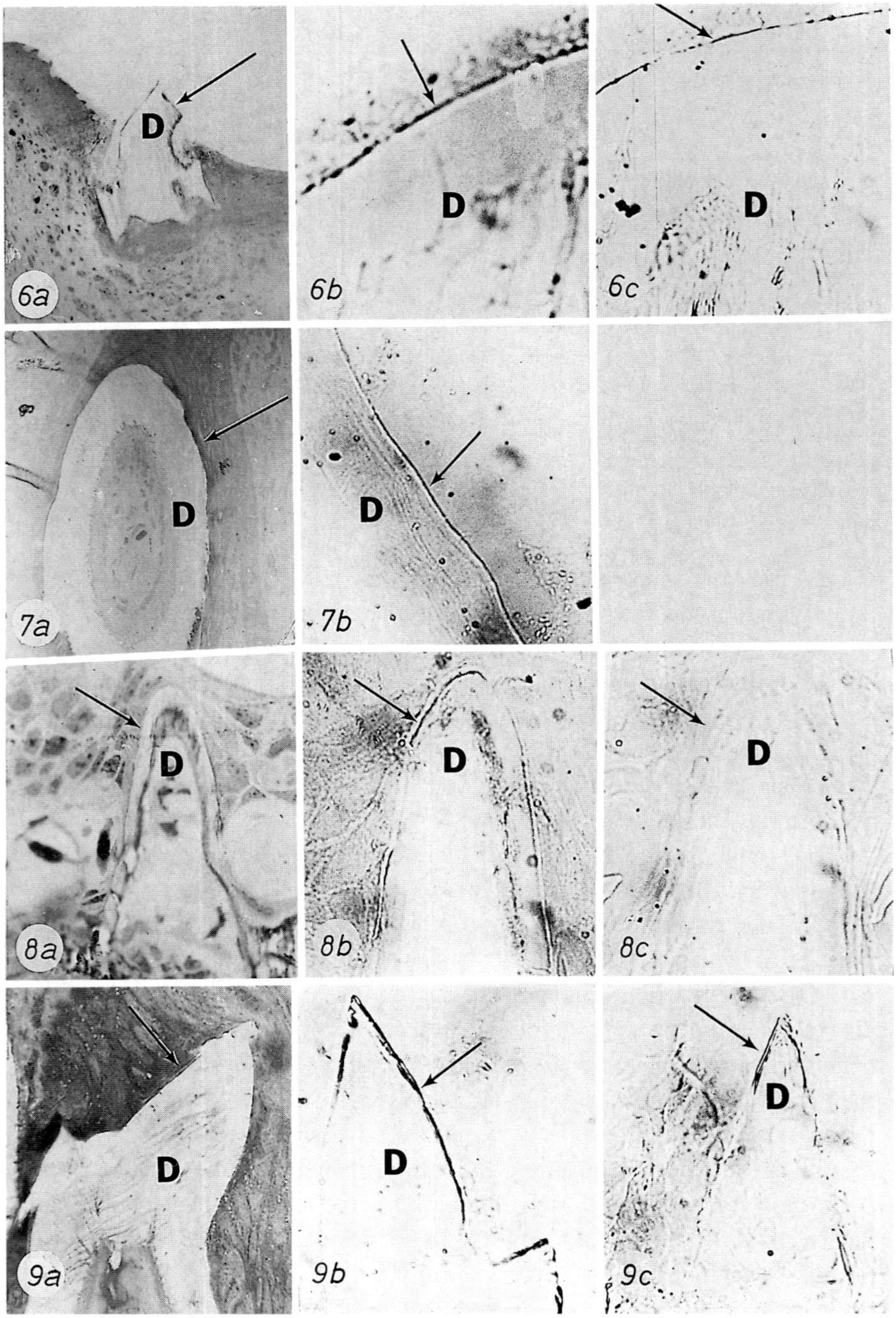


6

7

8

9

an acidic nonhistone chromosomal protein. Correlative evidence that it acts as a thyroid hormone receptor is that receptors are found only in responsive tissue; changes in receptor concentration correlate with changes in cell sensitivity to the hormone [51]. A proposed model suggests that the binding of thyroid hormone ($T_3$) to the receptor somehow alters gene expression, either by changing the chromatin conformation so that it is more accessible to RNA polymerases, or by catalyzing the action of other factors that are already stimulating the expression of particular DNA sequences [52–54]. Changes in gene expression in amphibian metamorphosis appear to be elicited by thyroid hormones. Although none of the above studies directly involved tooth tissues, the changes in expression of the genes for enamel proteins before and after amphibian metamorphosis suggest that they too may be controlled by the same mechanisms.

Genetic control of enamel formation in the salamander could work in the following way. The genome of the salamander contains the genes for both enamelin and amelogenin proteins. The inner enamel epithelial (IEE) cells of the tooth organs are the sites of expression of these genes. Teeth of two types are formed during salamander development: monocuspid larval teeth covered by enameloid are formed before metamorphosis and bicuspid adult teeth covered by enamel are formed after metamorphosis. The cells which make up the tooth-forming epithelial enamel organs for both the larval and

---

*Fig. 6.* Sections of adult bullfrog tooth. Arrows point to tooth-covering tissue. Tooth dentin (D): *(a)* stained with toluidine blue to show histology, ×64; *(b)* immunostained with antienamelin MAb immunogold-silver (note dark staining tooth-covering tissue, ×510); *(c)* immunostained with antiamelogenin MAb immunogold-silver (note dark staining tooth-covering tissue, ×510).

*Fig. 7.* Sections of neotenous mudpuppy tooth. Arrows indicate tooth-covering tissue. Tooth dentin (D): *(a)* stained with toluidine blue to show histology, ×64; *(b)* immunostained with antienamelin MAb immunogold-silver (note dark staining tooth-covering tissue, ×720).

*Fig. 8.* Sections of larval salamander tooth. Tooth-covering tissue (arrow) dentin (D): *(a)* stained with toluidine blue to show histology, ×255; *(b)* immunostained with anti-enamel in MAb immunogold-silver (note dark staining tooth-covering tissue, ×255); *(c)* immunostained with antiamelogenin MAb immunogold-silver (note lack of dark stain in tooth-covering tissue, ×255).

*Fig. 9.* Sections of adult salamander tooth. Tooth-covering tissue (arrow) dentin (D): *(a)* stained with toluidine blue to show histology, ×255; *(b)* immunostained with anti-enamelin MAb immunogold-silver (note dark stain of tooth-covering tissue, ×255); *(c)* immunostained with antiamelogenin MAb immunogold-silver (note dark staining of tooth-covering tissue, ×255).

the adult teeth are from the same group of cells, i.e. dental lamina. Only the gene for enamelin is expressed in the larvae IEE cells. During metamorphosis the increasing levels of thyroid hormone, utilizing the nuclear thyroid receptor in the IEE cells, activate the gene for amelogenins which are then produced in the cells and secreted along with enamelins into the extracellular matrix. The presence of these two proteins in the matrix results in an enamel-like tissue. In addition to this effect, metamorphosis also changes the shape of the tooth produced from monocuspid to bicuspid. The specific mechanisms involved in controlling and achieving this change of form are completely unknown.

## *References*


1 Snead AG, Lau EC, Zeichner-David M, et al: DNA sequence for cloned cDNA for murine amelogenin reveal the amino acid sequence for enamel-specific protein. Biochem Biophys Res Commun 1985;129:812–818.
2 Shimokawa H, Sobel ME, Sasaki ME, et al: Heterogeneity of amelogenin mRNA in the bovine tooth gene. J Biol Chem 1987;262:4042–4047.
3 Glimcher MJ, Friberg UA, Levine RT: The isolation and amino acid composition of the enamel proteins of erupted bovine teeth. Biochem J 1964;93:202–210.
4 Glimcher MJ, Mechanic GL, Bonar LC, et al: The amino acid composition of the organic matrix of decalcified fetal bovine dental enamel. J Biol Chem 1961;236: 3210–3213.
5 Glimcher MJ, Mechanic GL, Friberg UA: The amino acid composition of the organic matrix and the neutral-soluble and acid-soluble components of embryonic bovine enamel. Biochem J 1964;93:198–202.
6 Burgess RC, Maclaren CM: Proteins in developing bovine enamel; in Stack MV, Fernhead RW (eds): Tooth Enamel, Its Composition, Properties, and Fundamental Structure. Bristol, Wright, 1965, pp 74–82.
7 Eastoe JE: The amino acid composition of proteins from the oral tissues. II. The matrix proteins in dentine and enamel from developing human deciduous teeth. Arch Oral Biol 1963;8:633–652.
8 Katz EP, Mechanic GL, Glimcher MJ: The ultracentrifugal and free zone electrophoretic characterization of the neutral soluble proteins of embryonic bovine enamel. Biochim Biophys Acta 1965;107:471–484.
9 Levine PT, Glimcher MJ: The isolation and amino acid composition of the organic matrix and neutral soluble proteins of developing rodent enamel. Arch Oral Biol 1965;10:753–756.
10 Nikiforuk G, Gruca M: Immunological and gel-filtration characteristics of bovine enamel proteins; in Stack MV, Fearnhead RW (eds): Tooth Enamel, Its Composition, Properties, and Fundamental Structure. Bristol, Wright, 1971, vol 2, pp 95–99.
11 Termine JD, Belcourt AB, Christner PJ, et al: Properties of dissociatively extracted fetal tooth matrix proteins. J Biol Chem 1980;255:9760–9768.

12 Christner PJ, Lally ET, Ads EH, et al: Production of a monoclonal antibody to bovine tooth enamel proteins. Arch Oral Biol 1983;28:773–779.
13 Christner P, Lally E, Miller R, et al: Monoclonal antibodies to different epitopes in amelogenins from fetal bovine teeth recognize high-molecular weight components. Arch Oral Biol 1985;30:849–854.
14 Robinson C, Kirkham J: Enamel matrix components, alterations during development and possible interaction with the mineral phase; in Fearnhead RW, Sugs S (eds): Tooth Enamel, IVA. New York, Elsevier, 1984, pp 261–265.
15 Robinson C, Briggs HD, Kirkham J, et al: Changes in the protein components of rat incisor enamel during tooth development. Arch Oral Biol 1983;28:993–1000.
16 Rosenbloom J, Lally E, Dixon M, et al: Production of a monoclonal antibody to enamelins which does not cross-react with amelogenins. Calcif Tissue Int 1986;39: 412–415.
17 Graver HT, Herold RC, Chung TY, et al: Immunofluorescent localization of amelogenins in developing bovine teeth. Dev Biol 1978;63:390–401.
18 Herold R, Christner P, Graver H, et al: Immunofluorescent evidence for the similarity of amelogenins in calf, mouse and pig teeth. J Dent Res 1979;58B:992–993.
19 Sela M, Schechter B, Schechter I, Borek A: Antibodies to sequential and conformational determinants. Cold Spring Harbor Symp Quant Biol 1967;32:537–545.
20 Atassi MZ: Precise determination of the entire antigenic structure of lysozyme. Molecular features of protein antigenic structures and potential of 'surface-simulation' synthesis – a powerful new concept for protein binding sites. Immunochemistry 1978;15:909–936.
21 Benjamin DC, Berzofsky MA, East J, Gurd FRN, Hannum C, Leach SJ, Margoliash E, Michael JG, Miller A, Prager EM, Reichlin M, Sercarz EE, Smith-Gill SJ, Todd PE, Wilson AC: The antigenic structure of proteins: a reappraisal. Ann Rev Immunol 1984;2:67–101.
22 Schonfeld SE: Demonstration of an alloimmune response to embryonic enamel matrix proteins. J Dent Res 1975;54:C72–C77.
23 Schonfeld SE, Trump GN, Slavkin HC: Immunogenicity of two naturally occurring solid-phase enamel proteins. Proc Soc Exp Biol Med 1977;155:111–114.
24 Fincham AG, Belcourt AB, Lyaruu DM, et al: Comparative protein biochemistry of developing dental enamel matrix from five mammalian species. Calcif Tissue Int 1982;34:182–189.
25 Herold R, Boyde A, Rosenbloom J, et al: Monoclonal antibody and immunogold cytochemical localization of amelogenins in bovine secretory amelogenesis. Arch Oral Biol 1987;32:439–444.
26 Slavkin HC: Current perspectives on enamel proteins; in Butler W (ed): The Chemistry and Biology of Mineralized Connective Tissues. Birmingham, Alabama, Ebsco Media Inc, 1985, pp 237–239.
27 Slavkin HC, Zeichner-David M, MacDougall M, et al: Antibodies to murine amelogenins – localization of enamel proteins during tooth organ development in vitro. Differentiation 1982;23:73–82.
28 Slavkin HC, Zeichner-David M, Ferguson MWJ, et al: Phylogenetic and immunogenetic aspects of enamel proteins; in Rivere GR, Hildemann WH (eds): Oral Immunogenetics and Tissue Transplantation. Amsterdam, Elsevier, 1982, pp 241–251.
29 Zeichner-David M, MacDougall M, Slavkin HC: Enamelin gene expression during fetal and neonatal rabbit tooth organogenesis. Differentiation 1983;25:148–153.

30 Nanci A, Bendayan M, Slavkin HC: Enamel protein biosynthesis and secretion in mouse incisor secretory ameloblasts as revealed by high resolution immunocytochemistry. J Histochem Cytochem 1985;33:1153–1160.
31 Nanci A, Slavkin HC, Smith CE: Immunocytochemical and radioautographic evidence for secretion and intracellular degradation of enamel proteins by ameloblasts during the maturation stage of amelogenesis rat incisors. Anat Rec, 1987;217:107–123.
32 Slavkin HC, Zeichner-David M, MacDougall M, et al: Antibodies to murine amelogenins – localization of enamel proteins during tooth organ development in vitro. Differentiation 1982;23:73–84.
33 Robinson C, Kirkham J, Briggs HD, et al: Enamel proteins – from secretion to maturation. J Dent Res 1982;61:1490–1495.
34 Hayashi Y, Bianco P, Shimokawa H, et al: Organic-inorganic relationships and immunohistochemical localization of amelogenin and enamelins in developing enamel. Basic Appl Histochem 1986;30:291–299.
35 Slavkin H, Bessem C, Brigas P, et al: Sequential expression and differential function of multiple enamel proteins during fetal, neonatal, and early postnatal stages of mouse molar organogenesis. Differentiation 1988;37:26–39.
36 Zeichner-David M, Vides J, MacDougall M, et al: Biosynthesis and characterization of rabbit tooth enamel extracellular-matrix proteins. Biochem J 1988;251:631–641.
37 Herold R, Graver H, Christner PJ: Immunohistochemical localization of amelogenins in enameloid of lower vertebrate teeth. Science 1980;207:1357–1358.
38 Kemp NE: Organic matrices and mineral crystallites in vertebrate scales, teeth and skeletons. Am Zool 1984;24:965–976.
39 Slavkin HC, Graham EE, Zeichner-David M: Enamel-like antigens in hagfish – possible evolutionary significance. Evolution 1983;17:404–412.
40 Slavkin HC, Samuel N, Bringas P, et al: Selachian tooth development. II. Immunolocalization of amelogenin polypeptides in epithelium during secretory amelogenesis in *Squalus acanthias* J Craniofac Genet Dev Biol 1983;3:43–52.
41 Slavkin HC, Zeichner-David M, Snead ML, et al: Genetic and phylogenetic aspects of enamel; in Ruch JV, Belcourt A (eds): Tooth Morphogenesis and Differentiation. Paris, INSERM, 1984, pp 341–354.
42 Samuels N, Bessem C, Bringas P, et al: Immunochemical homology between elasmobranch scale and tooth extracellular matrix proteins in *Cephaloscylium ventriosum.* J Craniofac Genet Dev Biol 1987;7:371–388.
43 Herold R, Rosenbloom J, Granovsky M: Immunohistochemical localization of enamelin and amelogenin proteins in enamel and enameloid. Calcif Tissue Int 1987; 42(suppl 12):46P85.
44 Herold R, Rosenbloom J, Granovsky M: Phylogenetic distribution of enamel proteins: Immunohistochemical localization with monoclonal antibodies indicates the evolutionary appearance of enamelins prior to amelogenins. Calcif Tissue Int. 1989;45:88–94.
45 Kerr T: Development and structure of some actinopterygian and urodele teeth. Proc Zool Soc Lond 1958;133:401–422.
46 Clemen G, Greven H: Morphologische Untersuchungen an der Mundhöhle von Urodelen. III. Die Munddachbezahnung von *Ambystoma mexicanum.* Zool Jahrb Anat 1977:98:95–136.

47 Greven H, Clemen G: Morphological studies on the mouth cavity of urodeles. IV. The teeth of the upper jaw and palate in *Necturus maculosus*. Arch Hist Jap 1979;42:445–475.
48 Greven H, Laumeier I: A comparative SEM study on the teeth of 10 anuran species. Anat Anz Jena 1987;164:103–116.
49 Frieden E, Just J: Hormonal responses in amphibian metamorphosis; in Litwack G (ed): Biochemical Actions of Hormones. New York, Academic Press, 1970, vol I, pp 1–52.
50 Cohen P, Brucker R, Morris S: Cellular and molecular aspects of thyroid hormone action during amphibian metamorphosis; in Li C (ed): Hormonal Proteins and Peptides. Academic Press, New York, 1978, vol VI, pp 273–381.
51 Oppenheimer J, Surks M: Biochemical basis for thyroid hormone action; in Litwack G (ed): Biochemical Actions of Hormones. New York, Academic Press, 1975, vol III, pp 119–157.
52 Eberhardt N, Apriletti J, Baxtyer J: The molecular biology of thyroid hormone action; in Litwack G (ed): Biochemical Actions of Hormones. New York, Academic Press, 1980, vol VII, pp 311–394.
53 Oppenheimer J: Thyroid hormone actions at the cellular level. Science 1979;203: 971–979.
54 Chanoine C, d'Albis A, Lenfant-Guyot M, et al: Regulation by thyroid hormones of terminal differentiation in the skeletal dorsal muscle. II. Urodelan amphibians. Dev Biol 1987;123:33–42.
55 DeMey J, Hacker GW, De Waele M, Springall DR: Gold probes in light microscopy; in Polak J, Van Noorden S (eds): Immunocytochemistry. Amsterdam, Bristol, Wright, 1986, pp 71–88.

Richard Herold, MD, School of Dental Medicine, University of Pennsylvania, 4001 Spruce Street, Philadelphia, PA 19104 (USA)

Ferguson DB (ed): Aspects of Oral Molecular Biology.
Front Oral Physiol. Basel, Karger, 1991, vol 8, pp 42–56

# Molecular Changes in Dental Mesenchyme During Tooth Development

*Irma Thesleff*[a], *Markku Jalkanen*[b], *Anna-Maija Partanen*[a], *Seppo Vainio*[a]

[a]Department of Pedodontics and Orthodontics, University of Helsinki, and
[b]Department of Medical Biochemistry, University of Turku, Finland

## *Differentiation and Derivatives of the Dental Mesenchyme*

The dental mesenchyme differentiates from the neural crest-derived mesenchymal cells in the jaws. It is first morphologically identifiable as a zone of condensed cells which surrounds the epithelial tooth bud. The sequential differentiation of the dental mesenchymal cells results in the production of several highly specialized cell types (fig. 1). It is noteworthy that, with the exception of the enamel, which is an epithelial cell product, the major structures of the tooth and its supporting tissues are derivatives of the dental mesenchyme.

When the tooth undergoes morphogenesis from the bud to the cap stage of development, the dental mesenchyme segregates into two cell lineages, the dental papilla cells which are surrounded by the enamel epithelium, and the dental sac (follicle) cells which form the investing cell layer around the tooth germ. The dental papilla cells give rise to odontoblasts which differentiate from those cells which directly underline the epithelium, as well as to the fibroblasts of the tooth pulp [1]. It was recently suggested that the dental papilla would also be the source of periodontium [2], but the more general view is that it is the dental sac cells that differentiate into periodontal ligament fibroblasts, cementoblasts and osteoblasts, and form the periodontal tissues [3–5] (fig. 1).

As the various populations of dental mesenchymal cells reach higher levels of differentiation, their potential to express other differentiated phenotypes becomes restricted [3–5]. It is obvious that the sequential differentiation of the neural crest mesenchymal cells into the various highly differentiated cell types involves remarkable changes in the patterns of gene expression.

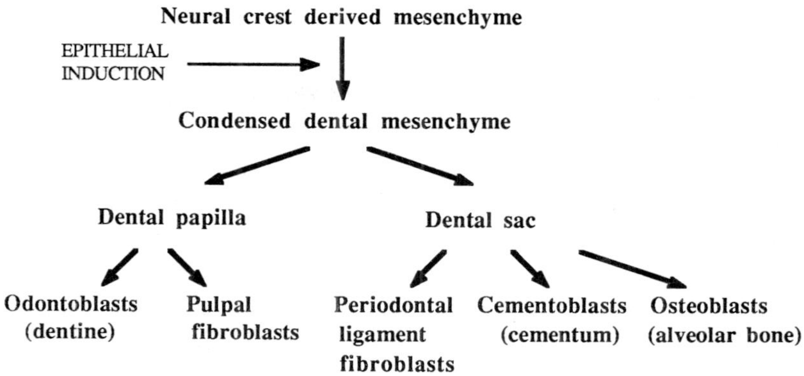


*Fig. 1.* Sequential differentiation of the dental mesenchymal cells. The presumptive dental epithelium induces the neural crest mesenchyme cells to form condensed dental mesenchyme. The subsequent development is controlled by reciprocal epithelial-mesenchymal interactions and results in morphogenesis and the differentiation of highly specialized cells forming the major structures of the tooth and its supporting tissues.

## *Regulation of Gene Expression in the Dental Mesenchyme*

The establishment of the lineage of the dental mesenchyme from the undifferentiated, neural crest-derived jaw mesenchyme is a result of epithelial-mesenchymal interaction. Experimental tissue recombination studies have shown that the presumptive dental epithelium in the frontonasal process and in the first branchial arch initially possesses the odontogenic potential, i.e. the capacity to form teeth [6]. This potential is shifted from the epithelium to the mesenchyme when the epithelium induces the condensation of the neural crest-derived jaw mesenchyme. Hence, if the dental mesenchyme is dissected from cap- or bell-staged tooth germs, and cultured with nondental epithelial tissue, it has the capacity to instruct enamel organ morphogenesis and ameloblast differentiation. The dental mesenchyme is also the only embryonic mesenchymal tissue that has the potential to differentiate into odontoblasts [1, 7]. It is clear that the sequential differentiation of the dental mesenchyme is intimately linked with the morphogenesis of the tooth and that the whole process is regulated by a series of reciprocal epithelial-mesenchymal interactions.

Of the various derivatives of the dental mesenchyme, the regulation of gene expression has been most extensively studied in the odontoblasts. The differentiation of the odontoblasts from the dental papilla mesenchyme is triggered by the enamel epithelium. Here, a cell-matrix interaction between the basement membrane and the mesenchymal cell surface has been shown to regulate gene expression [1, 8, 9]. More specifically, it has been suggested that the basement membrane contains time- and space-specific information which is recognized by receptors at the plasma membrane of mesenchymal cells and transduced and interpreted by the differentiating odontoblasts [10].

At present, our understanding of the molecular changes that accompany the sequential differentiation of the dental mesenchymal cells is limited. The gene expression of the terminally differentiated cells as well as the compositions of their specific products such as dentin, cementum and bone have been rather extensively investigated [11]. For example, osteonectin (SPARC) appears in mineralizing bone and dentine matrices [12] and its expression in osteoblasts and odontoblasts has been verified by in situ hybridization [13]. The expression of the dentine phosphoprotein appears to be unique for the odontoblasts [11, 14]. It is, however, apparent that molecular changes occur at even earlier developmental stages. The expression of odontogenic potential in the condensed dental mesenchymal cells must be reflected as molecular changes in the cells. Also, progressive differentiation as well as the restriction of developmental potential of the cells in the various lineages must result from differential gene expression.

In this review, we summarize our findings on the changes of expression of some molecules, which appear to be associated with the advancing differentiation of the dental mesenchyme and with tooth morphogenesis and which also appear to be regulated by epithelial-mesenchymal interactions.

## *Results and Discussion*

### *Localization of Binding Sites for Epidermal Growth Factor (EGF)*

Growth factors are polypeptides which affect cell proliferation and also other cellular functions by binding to specific cell surface receptors. This implies that only cells which express these receptors are responsive to the growth factors. The receptor molecules are transmembrane proteins, which are often tyrosine-specific protein kinases and are activated by binding of the growth factor ligand to the extracellular portion of the receptor molecule [15]. Several growth factors and their receptors have been detected in embryonic tissues and it is apparent that growth factors function locally in an autocrine or paracrine manner [16–18].

The presence of EGF-like activity and appearance of EGF receptors in embryonic tissues indicate that EGF and/or transforming growth factor α, which competes for EGF receptors and may represent the embryonic form of EGF, play physiological roles during embryonic development [17, 18]. EGF causes precocious tooth eruption in newborn mice [19] and it inhibits morphogenesis of mouse embryonic cap-staged teeth in organ culture [20].

We have localized the binding sites of EGF at various stages of tooth development to examine how the responsiveness to EGF is related to the presence and distribution of receptors. Localization of EGF binding by using $^{125}$I-labeled EGF and autoradiography of tissue sections revealed that distribution of receptors in dental tissues, and also expression of receptors in the dental mesenchyme, changes markedly from bud stage to bell stage of development [21]. At bud stage, there are no receptors in the condensed dental mesenchyme, while the dental epithelial bud binds EGF (fig. 2a). At cap stage, EGF receptors have appeared in the dental mesenchyme, but binding in the dental epithelium has decreased (fig. 2b). As development proceeds from cap stage to bell stage, binding of EGF decreases in the dental papilla mesenchyme (fig. 2c).

Expression of EGF receptors in dental mesenchyme is clearly associated with the stage of active growth during tooth crown morphogenesis, and is decreased with the reduction of cell proliferation and the onset of odontoblast and ameloblast differentiation. However, the stimulation of cell proliferation in tooth germs by EGF is not correlated with the distribution of EGF receptors. Hence, the response of dental tissues to EGF may be controlled by epithelial-mesenchymal tissue interactions, where the dental mesenchyme is the primary target tissue and mediates the stimulatory effect to the dental epithelium.

At the bell stage, when the mesenchymal cells in the dental papilla no longer bind EGF, the cells in the dental sac show abundant binding (fig. 2c) which is still evident in erupting teeth [22]. The intense EGF binding in the lineage of dental sac (follicle) cells may be associated with the responsiveness of periodontal tissues to EGF-like growth factors during tooth eruption [23].

### *Expression of the Matrix Protein Tenascin by the Dental Mesenchyme*

Tenascin, which is also known as cytotactin, J1 glycoprotein and GMEM, is a large extracellular matrix glycoprotein consisting of six disulfide-linked subunits of 190–240 × $10^3$ $M_r$. It is present in restricted locations in the embryo, but largely absent from mature connective tissues [25]. Tenascin already appears in the dental mesenchyme when it is first morpho-

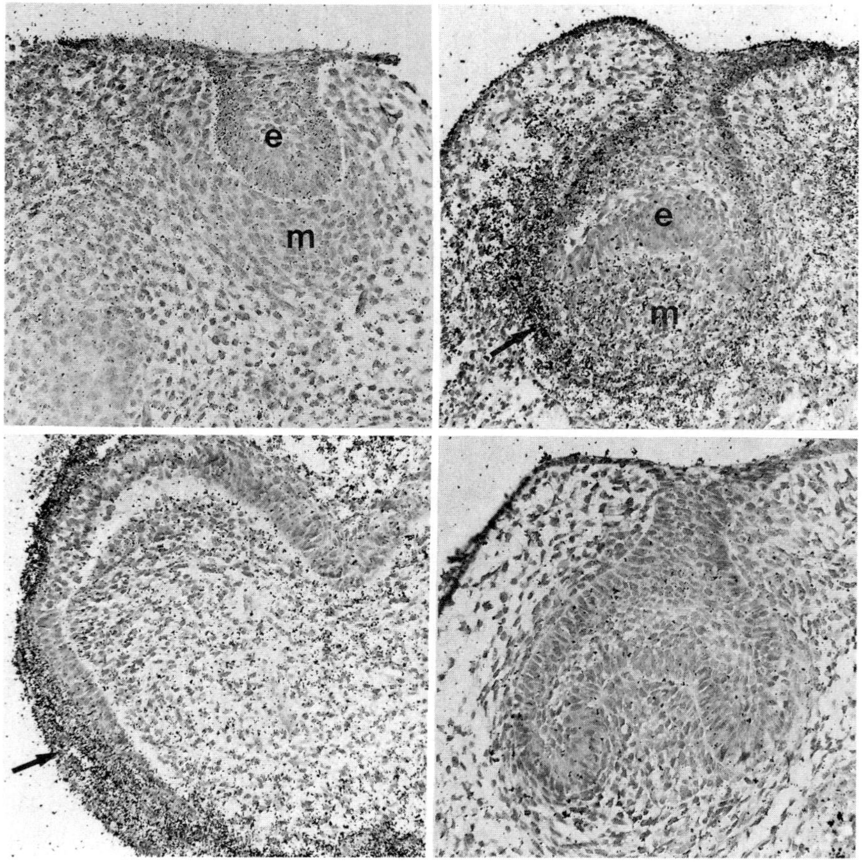


*Fig. 2.* Autoradiographs showing the distribution of $^{125}$I-labeled EGF binding in mouse embryonic first mandibular molar tooth germs at the bud, cap and bell stages of development. [From 24, reproduced with permission.] *a* In the bud-staged tooth, the condensed dental mesenchyme (m) is sparsely labeled, whereas the epithelial bud (e) and the nondental mesenchyme around the tooth germ are moderately labeled. *b* At the early cap stage, the cells of the dental mesenchyme both in the dental papilla (m) and dental sac (arrow) are labeled but in the dental epithelium (e) only the outermost cells have bound $^{125}$I-EGF. The surrounding nondental mesenchyme is also labeled. *c* At the bell stage the label is very sparse in the dental mesenchyme and in the dental epithelium, while the cells of the dental sac or follicle (arrow) surrounding the tooth germ are densely labeled. *d* Autoradiograph of day 14 tooth incubated in the presence of 5 μg/ml of unlabeled EGF. The labeling by $^{125}$I-EGF is totally abolished, which confirms the specificity of label in figure 2a–c.

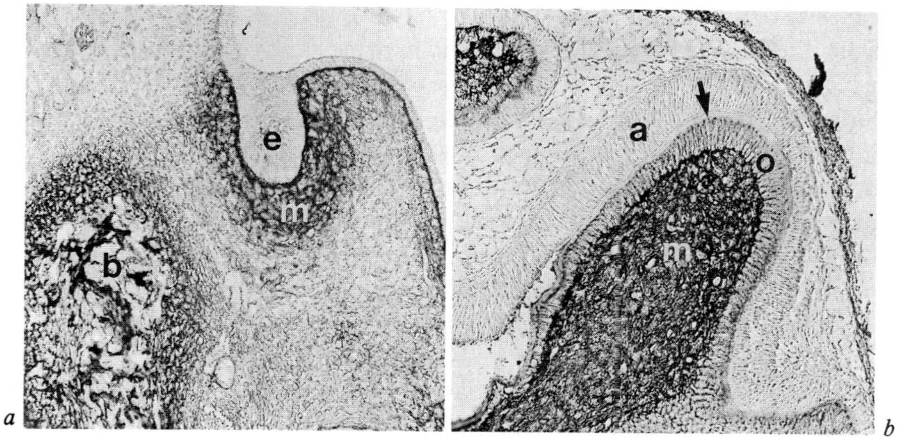


*Fig. 3.* Immunoperoxidase localization of the matrix protein tenascin in mouse tooth germs by a polyclonal rabbit antibody. *a* At the bud stage intense staining is seen in the condensing dental mesenchyme around the epithelial bud and in the osteogenic mesenchyme of the forming mandibular bone. *b* At the late bell stage (newborn mouse molar) intense staining has remained in the dental mesenchyme. Some staining is seen between the odontoblasts but dentin matrix is negative (arrow). e = Epithelium; m = dental mesenchyme; a = ameloblasts; o = odontoblasts; b = bone.

logically identifiable. It is strikingly accumulated around the condensing mesenchymal cells in the bud-staged tooth germ (fig. 3a) [25, 26]. This restricted pattern is unlike that of other matrix molecules such as fibronectin and collagens which are evenly distributed throughout jaw mesenchyme [27]. Besides condensing dental mesenchyme, tenascin is intensely expressed by the early organ-specific mesenchyme also in the developing mammary gland and hair follicle [25]. This suggests that tenascin is involved in mesenchymal cell condensation during early organ morphogenesis.

Tenascin remains intensely expressed in the dental papilla mesenchyme throughout tooth development and it is still present in the pulp of erupted teeth (fig. 3b) [26, 28]. The dental sac mesenchyme at the bell stage appears, however, negative for tenascin. Tenascin is present in the dental basement membrane at the time of odontoblast differentiation, but differentiated odontoblasts and the dentine matrix are negative (fig. 3b) [26, 28].

Tenascin is also absent from mature cartilage and bone, but it is intensely expressed during early chondrogenesis and osteogenesis (fig. 3a) [29]. Since tenascin is expressed particularly by cells in the osteoblastic, chondroblastic and odontoblastic cell lineages, it may be associated with the capacity of differentiation into hard tissue-forming cells. Hence, the expression of tenascin in the dental papilla and pulp mesenchyme throughout tooth development may be associated with the capacity of cells in this lineage to differentiate, when suitably triggered, into odontoblasts. In this respect, it is of interest that tenascin was recently shown to interfere with fibronectin action both by affecting receptor-mediated cell attachment and by directly binding to the fibronectin molecule [30], and that fibronectin has been associated with the induction of odontoblast differentiation earlier [1, 7, 10]. Hence, it is possible that an interaction between fibronectin and tenascin is involved when the mesenchymal cells polarize and differentiate into odontoblasts.

*Transient Expression of a Cell Surface Proteoglycan in Dental Mesenchyme*

Like growth factors, extracellular matrix molecules also exert their effects on cells at the cell surface. Examples of cell surface matrix receptors are the integrins which recognize RGD sequences in fibronectin and other molecules [31]. We have studied the expression of a cell surface proteoglycan (PG) which was originally isolated from epithelial cells and which also functions as a matrix receptor [32, 33, 50]. The ectodomain of this heparan sulfate-rich PG binds to interstitial collagens [34] and fibronectin [35]. Because it associates intracellularly with cytoskeletal elements when cross-linked at the cell surface, it is believed that this PG is involved in the transmission of changes in the extracellular matrix into changes of cellular behavior [33, 36].

We have used the monoclonal antibody (MAb) 281-2 [37], which recognizes the extracellular domain of the core protein of this PG, for immunohistochemical localization studies during early tooth development. Our recent observations indicate that the cell surface PG, which is expressed exclusively by epithelial cells in adult tissues [38], is transiently expressed by embryonic dental mesenchyme [39]. The PG is very intensely expressed in the condensing dental mesenchyme of the bud-staged tooth (fig. 4a). At the cap stage, the staining remains intense in the dental papilla, as well as in the dental sac mesenchyme, but during bell stage the staining is lost from most of the dental papilla, and the odontoblasts are completely negative. Weak expression of the PG persists in the dental sac mesenchyme (fig. 4c).

The possibility that the PG in the dental mesenchyme would have originated from the epithelium by proteolytic cleavage of the ectodomain [40] was excluded by metabolic labeling of the isolated tissue components. The epithelia of cap-staged molars from 14-day-old mouse embryos were separated and the dental mesenchyme was dissected free from the surround-

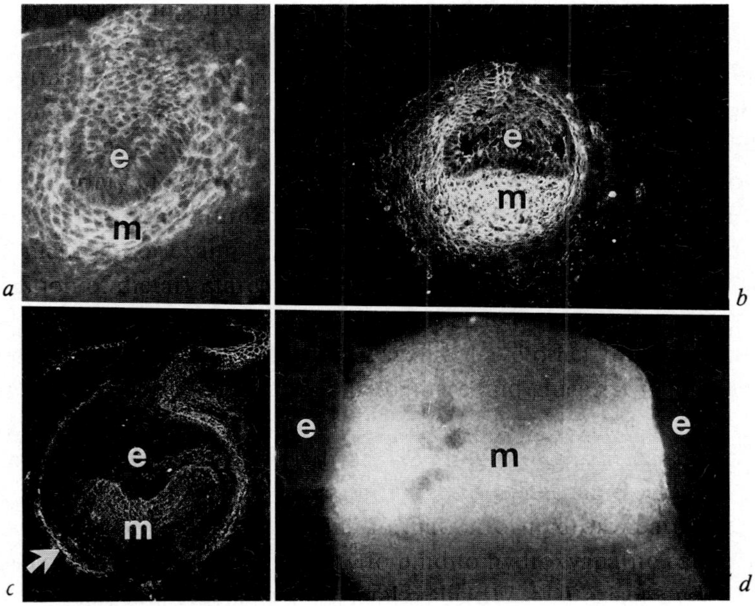


*Fig. 4.* Immunofluorescent localization of cell surface proteoglycan with the mouse-specific MAb 281-2. *a, b* In bud- and cap-staged mouse tooth germs the dental mesenchyme is intensely stained, whereas the surrounding jaw mesenchyme is negative. The oral epithelium is stained but the dental epithelium loses stain with advancing development. *c* At the bell stage staining in the dental papilla mesenchyme has dramatically decreased, but staining persists in the dental sac cells (arrow). *d* An experimental recombinant of two pieces of rat presumptive dental epithelia (not stained with MAb 281-2) cultured in contact with mouse presumptive dental mesenchyme. The epithelium has induced PG expression in a restricted area of the mesenchyme. e = Dental epithelium; m = dental mesenchyme.

ing jaw mesenchyme. It was shown that, in addition to the epithelium, the dental mesenchyme, but not the surrounding jaw mesenchyme, synthesized the PG (fig. 5) [41].

The appearance of the PG in the early condensing dental mesenchyme and its disappearance by the time of completed tooth morphogenesis suggests that it has an important developmental function. Also, the shift of PG expression from the dental epithelium to the condensing dental mesenchyme at the same developmental stage when the odontogenic potential is transferred, may indicate that this PG is involved in the determination of the dental mesenchyme from the neural crest-derived jaw mesenchyme.

*Expression of Tenascin and Cell Surface PG in Dental Mesenchyme Is Regulated by Epithelial-Mesenchymal Interaction*

The role of the epithelium in the regulation of tenascin and PG expression was studied in in vitro tissue recombinants. The presumptive dental epithelium, which still has the odontogenic potential, was isolated from 11-day-old mouse or 13-day-old rat embryos, and recombined with the underlying jaw mesenchyme. In such recombinants, PG and tenascin expression in the mesenchyme was shown to be induced by epithelium [42]. The staining appeared first at the epithelial-mesenchymal interface and then spread through the mesenchyme during increased culture time (fig. 4d). Tenascin and PG staining localized to a zone of reorganized mesenchyme which appeared as a bright area in phase-contrast microscopy. This area probably represents a mass of similarly determined cells which are active during organ morphogenesis [43]. When the mesenchyme was cultured without epithelial contact, tenascin positivity was not detected and PG staining in the mesenchyme was weak and homogeneous.

Our results show that in the presumptive dental mesenchyme, PG as well as tenascin expression is regulated by epithelial-mesenchymal interactions. The co-localization of PG and tenascin in the in vitro combinants as well as in vivo

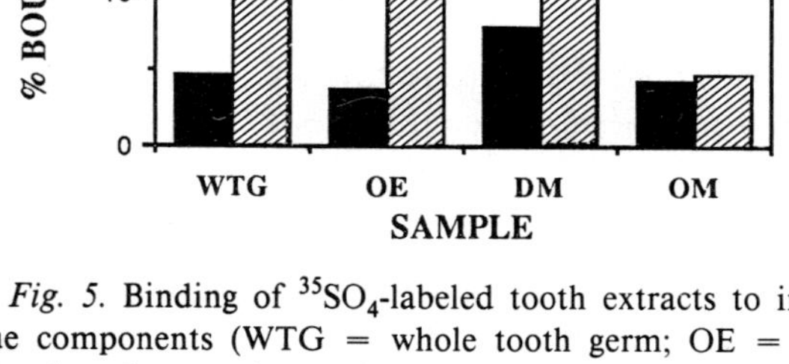


*Fig. 5.* Binding of $^{35}SO_4$-labeled tooth extracts to immunoaffinity columns. The tissue components (WTG = whole tooth germ; OE = oral and dental epithelium; DM = dental mesenchyme; OM = oral mesenchyme) from cap-staged teeth were separated and labeled with radiosulfate in separate cultures. Tissue extracts were then eluted both through Hermes-1 (negative control) and 281-2 affinity columns. Figures represent percent of total PG-bound radioactivity retarded by Hermes-1 (■) and 281-2 (▨) column. For each tissue extract the mean of three separate isolations are indicated.

in the condensed dental mesenchyme suggests that these molecules may interact. Therefore, it is possible that the PG and tenascin represent a cell surface receptor and its matrix ligand, which are involved in the establishment of odontogenic pattern and condensation of the dental mesenchyme.

*Recognition of a Developmentally Regulated Antigen in the Dental Mesenchyme by a Monoclonal Antibody*

We have used hybridoma technology in order to develop differentiation markers for the dental mesenchyme and to identify some molecular specificities of this unique tissue [44].

Dental papillae from bell-staged embryonic molars were separated from the epithelium and gently homogenized. This material was used for immunization of rats and mice. Hybridomas were produced by standard techniques and screened by immunohistology. One hybridoma, 7C5, stained intensely the jaw mesenchyme in 11-day-old mouse embryos. During bud and cap stages of tooth development, the staining became more restricted to the condensed dental mesenchyme. During bell stage, the dental papilla, i.e. the tissue which was used as the immunogen, was stained particularly intensely with MAb 7C5 whereas the surrounding nondental mesenchyme was largely negative (fig. 6a). The staining was lost from the polarized mesenchymal cells during terminal differentiation of the odontoblasts (fig. 6b).

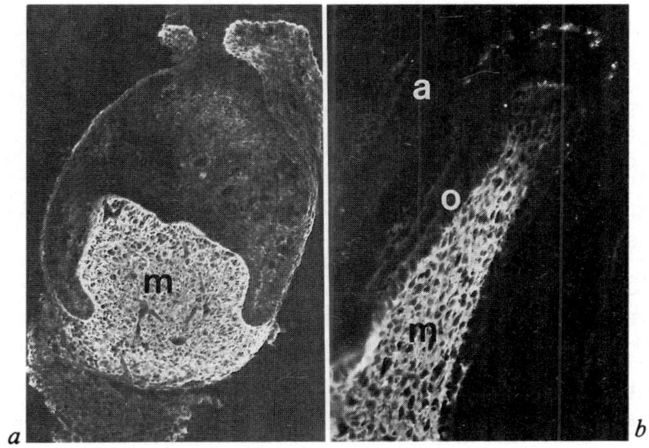


*Fig. 6.* Immunofluorescent staining of paraffin sections through an early *(a)* and late bell-staged mouse molar *(b)* with the MAb 7C5. Homogenized dental papillae were used as immunogen in producing this MAb. The dental mesenchyme (m) is intensely stained, but the differentiated odontoblasts (o) are negative. a = Ameloblasts.

The distribution pattern of the antigen identified with MAb 7C5 does not resemble that of other molecules so far localized in dental tissues. Hence it is possible that we have discovered a new developmentally regulated antigen. However, the MAb 7C5 also stains various other embryonic tissues. Whether the dental papilla does, in fact, express molecules specific only to this embryonic cell population remains to be clarified.

## *Conclusions and Hypotheses*

Our observations indicate that the acquisition of higher levels of determination and differentiation of the dental mesenchymal cells is associated with multiple molecular changes. Changes in the binding affinity of EGF as well as the changed expression patterns of the cell surface PG which binds interstitial matrix molecules, indicate that progressive differentiation is accompanied by alterations in the responsiveness of the cells to growth factors and matrix molecules in their environment.

Through binding to their cell surface receptors, both growth factors and extracellular matrix molecules affect a variety of cellular functions. It has also become evident that the actions of growth factors and matrix molecules are closely linked. For instance, the extracellular matrix influences the responses of cells to growth factors and functions in their anchorage [45]. Growth factor-like sequences have been detected in some matrix molecules [46]. Finally, many growth factors specifically influence the turnover of the extracellular matrix. An interesting growth factor in this respect is transforming growth factor β (TGFβ) which markedly enhances matrix accumulation. It was recently shown that TGFβ is intensely expressed in embryonic mesenchymal tissues, including dental mesenchyme, and it was suggested that TGFβ could be involved in the induction of osteonectin (SPARC) expression by the odontoblasts [47].

The transient expression of the cell surface PG by the dental mesenchymal cells between the initiation of condensation and the bell stage is closely associated with morphogenesis. The PG may be involved in transmission of signals in the extracellular matrix to induce mesenchymal cell behavior. Since the PG bears predominantly heparan sulfate GAG chains, it may also be involved in anchorage of growth factors. In this respect, the partial co-distribution of EGF-binding sites and PG expression may be significant.

The matrix protein tenascin and the cell surface PG are early molecular markers of the differentiation of neural crest cells into dental mesenchymal cells. Our observation that tenascin and PG appear in the condensing dental mesenchyme as a result of epithelial-mesenchymal interaction may indicate a role for them in the early determination of the lineage of dental mesenchymal cells. Tenascin was expressed throughout tooth development by the dental mesenchymal cells that form the dental papilla and later persist as fibroblasts in the dental pulp. Hence, tenascin expression may be associated with the capacity of the dental mesenchymal cells to differentiate into odontoblasts.

By raising monoclonal antibodies against the dental papilla mesenchyme we have identified molecules which are intensely expressed by dental mesenchymal cells, but so far we have not found specific differentiation markers. Whether the specificities of an ancestral cell lineage such as the dental mesenchyme, and of its derivatives are determined by the expression of unique lineage-specific molecules, or by unique combinations, or balances, of several molecules, remains to be shown. However, hybridoma technology offers a tool to detect developmentally significant molecules, the expression of which may be limited to specific cell populations.

## *References*

1 Thesleff I, Hurmerinta K: Tissue interactions in tooth development. Differentiation 1981;18:75–88.
2 Osborn JW, Price DC: An autoradiography study of periodontal development in the mouse. J Dent Res 1988;67:455–461.
3 Ten Cate AR, Mills C, Solomon G: The development of the periodontium. A transplantation and autoradiographic study. Anat Rec 1981;170:365–379.
4 Yoshikawa DK, Kollar EJ: Recombination experiments on the odontogenic roles of mouse dental papilla and dental sac tissues in ocular grafts. Arch Oral Biol 1981;26: 303–307.
5 Palmer RM, Lumsden AGS: Development of periodontal ligament and alveolar bone in homografted recombinations of enamel organs and papillary, pulpal and follicular mesenchyme in the mouse. Arch Oral Biol 1987;32:281–289.
6 Mina M, Kollar EJ: The induction of odontogenesis in non-dental mesenchyme combined with early murine mandibular arch epithelium. Arch Oral Biol 1987;32: 123–127.
7 Kollar EJ, Baird G: Tissue interactions in developing mouse tooth germs. II. The inductive role of the dental papilla. J Embryol Exp Morphol 1970;24:173–186.
8 Thesleff I, Lehtonen E, Saxén L: Basement membrane formation in transfilter tooth culture and its relation to odontoblast differentiation. Differentiation 1978;10:71–79.
9 Ruch JV, Lesot H, Karcher-Djuricic V, et al: Facts and hypotheses concerning the control of odontoblast differentiation. Differentiation 1982;21:7–12.

10 Ruch JV: Determinisms of odontogenesis. Cell Biol Rev RBC 1987;14:1–112.
11 Butler WT: Matrix macromolecules of bone and dentin. Coll Relat Res 1984;4:297–307.
12 Termine JD, Kleinman HK, Whitson SW, et al: Osteonectin, a bone-specific protein linking mineral to collagen. Cell 1981;26:99–105.
13 Holland PWH, Harper SJ, McVey JH, et al: In vivo expression of mRNA for the $Ca^{++}$-binding protein SPARC (osteonectin) revealed by in situ hybridization. J Cell Biol 1987;105:473–482.
14 Rahima M, Veis A: Two classes of dentin phosphophoryns, from a wide range of species, contain immunologically cross-reactive epitope regions. Calcif Tissue Int 1988;42:104–112.
15 Hunter T: A thousand and one protein kinases. Cell 1987;50:823–829.
16 Mercola M, Stiles CD: Growth factor superfamilies and mammalian embryogenesis. Development 1988;102:451–460.
17 Adamson ED, Deller MJ, Warshaw JB: Functional EGF receptors are present on mouse embryo tissues. Nature (Lond) 1981;291:656–659.
18 Popliker M, Shatz A, Avivi A, et al: Onset of endogenous synthesis of epidermal growth factor in neonatal mice. Dev Biol 1987;119:38–44.
19 Cohen S: Isolation of a mouse submaxillary gland protein accelerating incisor eruption and eyelid opening in the newborn animal. J Biol Chem 1962;237:1555–1562.
20 Partanen A-M, Ekblom P, Thesleff I: Epidermal growth factor inhibits morphogenesis and cell differentiation in cultured mouse embryonic teeth. Dev Biol 1985;111:84–94.
21 Partanen A-M, Thesleff I: Localization and quantitation of $^{125}$I-epidermal growth factor binding in mouse embryonic tooth and other embryonic tissues at different developmental stages. Dev Biol 1987;120:186–197.
22 Thesleff I, Partanen A-M, Rihtniemi L: Localization of epidermal growth factor receptors in mouse incisors and human premolars during eruption. Eur J Orthod 1987;9:24–32.
23 Thesleff I: Does epidermal growth factor control tooth eruption? J Dent Child 1987;54:321–329.
24 Partanen A-M: Epidermal growth factor and tooth development. Proc Finn Dent Soc 1987;83:181–190.
25 Chiquet-Ehrismann R, Mackie EJ, Pearson CA, et al: Tenascin: An extracellular matrix protein involved in tissue interactions during fetal development and oncogenesis. Cell 1986;47:131–139.
26 Thesleff I, Mackie E, Vainio S, et al: Changes in the distribution of tenascin during tooth development. Development 1987;101:289–296.
27 Thesleff I, Stenman S, Vaheri A, et al: Changes in the matrix proteins, fibronectin and collagen, during differentiation of mouse tooth germ. Dev Biol 1979;70:116–126.
28 Thesleff I, Kantomaa T, Mackie E, et al: Immunohistochemical localization of the matrix glycoprotein tenascin in the skull of the growing rat. Arch Oral Biol 1988;33:383–390.
29 Mackie EJ, Thesleff I, Chiquet-Ehrismann R: Tenascin is associated with chondrogenic and osteogenic differentiation in vivo and promotes chondrogenesis in vitro. J Cell Biol 1987;105:2569–2579.

30 Chiquet-Ehrismann R, Kalla P, Pearson CA, et al: Tenascin interferes with fibronectin action. Cell 1988;53:383–390.
31 Ruoslahti E, Pierschbacher M: New perspectives in cell adhesion. Science 1987;238:491–497.
32 Rapraeger AC, Bernfield MR: Heparan sulfate proteoglycans from mouse mammary epithelial cells. A putative membrane proteoglycan associates quantitatively with lipid vesicles. J Biol Chem 1983;258:3632–3636.
33 Rapraeger AC, Jalkanen M, Bernfield MR: Cell surface proteoglycan associates with the cytoskeleton at the basolateral cell surface of mouse mammary epithelial cells. J Cell Biol 1986;103:2683–2696.
34 Koda JE, Rapraeger A, Bernfield M: Heparan sulfate proteoglycans from mouse mammary epithelial cells: cell surface proteoglycan as a receptor for interstitial collagens. J Biol Chem 1985;260:8157–8162.
35 Saunders S, Bernfield MR: Cell surface proteoglycan binds mouse mammary epithelial cells to fibronectin and behaves as an interstitial matrix receptor. J Cell Biol 1988;106:423–430.
36 Jalkanen M: Biology of cell surface heparan sulfate proteoglycans. Med Biol 1987;65:41–47.
37 Jalkanen M, Nguyen H, Rapraeger A, et al: Heparan sulfate proteoglycans from mouse mammary epithelial cells: Localization on the cell surface with a monoclonal antibody. J Cell Biol 1985;101:976–984.
38 Hayashi K, Hayashi M, Jalkanen M, et al: Immunocytochemistry of cell surface heparan sulfate proteoglycan. A light and electron microscopic study. J Histochem Cytochem 1987;35:1079–1088.
39 Thesleff I, Jalkanen M, Vainio S, et al: Cell surface proteoglycan expression correlates with epithelial-mesenchymal interactions during tooth morphogenesis. Dev Biol 1988;129:565–572.
40 Jalkanen M, Rapraeger A, Saunders S, Bernfield M: Cell surface proteoglycan of mouse mammary epithelial cells is shed by cleavage of its matrix-binding ectodomain from its membrane-associated domain. J Cell Biol 1987;105:3087–3096.
41 Salmivirta M, Vainio S, Elenius K, Chiquet-Ehrismann R, Thesleff I, Jalkanen M: Cell surface proteoglycan, syndecan, from embryonic tooth mesenchyme binds matrix glycoprotein, tenascin (submitted).
42 Vainio S, Jalkanen M, Thesleff I: Syndecan and tenascin expression is induced by epithelial-mesenchymal interactions in embryonic tooth mesenchyme. J Cell Biol 1989;108:1945–1954.
43 Saxén L, Karkinen-Jääskeläinen M, Lehtonen E, et al: Inductive tissue interactions; in Poste G, Nicolson GL (eds): The Cell Surface in Animal Embryogenesis and Development. Amsterdam, Elsevier/North Holland, 1976, pp 331–407.
44 Vainio S, Lehtonen E, Kaartinen M, et al: Production of monoclonal antibodies against murine dental papilla. Scand J Dent Res 1988;96:177–187.
45 Gospodarowicz D, Neufeld G, Schweigerer L: Molecular and biological characterization of fibroblast growth factor, an angiogenic factor which also controls the proliferation and differentiation of mesoderm and neuroectoderm derived cells. Cell Differ 1986;19:1–17.
46 Krusius T, Gehlsen KR, Ruoslahti E: A fibroblast chondroitin sulfate proteoglycan core protein contains lectin-like and growth factor-like sequences. J Biol Chem 1987;262:13120–13125.

47 Heine UI, Monoz EF, Flanders KC, et al: Role of transforming growth factor-β in the development of the mouse embryo. J Cell Biol 1987;105:2861–2876.
50 Saunders S, Jalkanen M, O'Farrell S, Bernfield M: Molecular cloning of syndecan, an integral membrane proteoglycan. J Cell Biol 1989;108:1547–1556.

## *Note Added in Proof*

After the completion of this manuscript, the cell surface proteoglycan (PG) has been cloned and named syndecan.

Dr. Odont. Irma Thesleff, DDS, Institute of Dentistry, University of Helsinki, Mannerheimintie 172, SF–00300 Helsinki (Finland)

Ferguson DB (ed): Aspects of Oral Molecular Biology.
Front Oral Physiol. Basel, Karger, 1991, vol 8, pp 57–76

# Genetic Analysis and Genetic Probes for Oral Bacteria

*R.R.B. Russell*

Hunterian Dental Research Unit, London Hospital Medical College, London, UK

The two most prevalent oral diseases, dental caries and periodontal disease, have long been known to be associated with the microbial component of dental plaque. Indeed, it is now over a century since the significance of bacteria in initiating tooth decay was first demonstrated. The continuing difficulty in obtaining a full understanding of the disease processes however, is encapsulated in the debate on the relative merits of the specific versus the nonspecific plaque hypothesis. In other words, the question is as to whether there are particular aetiological organisms or whether the entire complex community of bacteria resident in plaque is responsible. As is usual in most controversies, the truth lies somewhere between the extremes but it is the very complexity of dental plaque composition which has precluded the acquisition of clear answers. Therefore, if we wish to discover more about the causation of caries and the range of afflictions collectively referred to as periodontal diseases, and ultimately devise rational strategies for prevention, prognosis or treatment, it is necessary to attack the question at a number of different levels – (a) develop methods for the precise identification of all bacteria present in plaque; (b) determine the association of particular bacteria with disease status; (c) carry out detailed investigation of the mechanism of pathogenesis. Although very considerable advances have been made in each of these areas of oral microbiology, particularly in the last 15–20 years, much remains to be learnt and the application of molecular biology techniques is making its impact and quickening the pace of discovery. Before considering the application of the new technology, let us briefly review the main problems to be addressed.

## *Dental Plaque*

Right from the moment of birth the oral cavity is colonized by bacteria, initially by species such as *Streptococcus salivarius* which have specificity for mucosal surfaces but, as soon as teeth begin to erupt, species able to colonize hard surfaces can be found. Species of *Streptococcus* and *Actinomyces* are pioneers in colonization of the teeth, but there is a gradual increase in complexity of the microbial population so that well-established dental plaque may yield representatives of several hundred different species from a wide range of genera. The maintenance of this highly complex microbial ecosystem is dependent upon a huge array of interactions between bacteria and between bacteria and the host. Some of these interactions are highly specific, for example binding reactions between particular bacteria and salivary macromolecules, between different species of bacteria (in coaggregation pairs) or between bacteria and extracellular polymers such as dextran, which are synthesized in plaque. Other interactions do not manifest a high level of specificity and any of a number of species may benefit (or suffer) from the activities of others. Examples of negative interactions include the production of antagonistic substances such as bacteriocins or hydrogen peroxide, while positive effects are seen in the occurrence of mutually beneficial consortia of organisms which act together to break down macromolecules into assimilable form. This latter topic is attracting increasing attention as it is now appreciated that the major source of nutrients for supragingival plaque bacteria are glycoproteins from saliva, since multiplication of plaque bacteria is sustained in the absence of any dietary intake. In subgingival plaque, where saliva and host diet have no access, it is macromolecules from periodontal tissues and serum which are utilized. These differing nutritional sources, and the anaerobiosis of conditions below the gingival margin, contribute to the substantial variation in the composition of the microbial populations of supra- and subgingival plaques.

## *Association of Bacteria with Oral Diseases*

### *Dental Caries*

Evidence on the identity of bacteria associated with the initiation and progression of dental caries comes from both cross-sectional and longitudinal epidemiological studies in human populations, supported by results from experimental animal studies and in vitro characterization of suspected cariogenic organisms. The review by Loesche [1] summarizes the research which has led to the recognition of the importance of mutans streptococci and, less conclusively, the involvement of lactobacilli. The mutans streptococci illustrate well some of the difficulties in identification, taxonomy and nomenclature of oral bacteria. Despite the early efforts of Clarke, who first recognized an association between *Streptococcus mutans* and dental caries in 1924, right up until the late 1960s, this species was lost in the heterogeneous group of organisms referred to as the viridans streptococci. This group is now known to contain a dozen or more distinct species, though uncertainties still remain about the precise relationship of strains known as *Streptococcus oralis* and *Streptococcus mitior*, the true number of species within the *Streptococcus milleri* complex and several other areas [2]. These difficulties arise because of the paucity of easily determined phenotypic characters amongst the streptococci and, because of the lack of physiological or serological tests for discrimination, DNA/DNA hybridization studies have been crucial in resolving the taxonomic relationships. First applied to oral streptococci by Coykendall [3], this approach is based upon the fact that if chromosomal DNA from two separate bacterial isolates can be shown to be greater than 70% homologous (i.e. more than 70% of their nucleotide sequences are identical) by experiments which measure the extent of reassociation of denatured single strands of genomic DNA, then they are considered to belong to a single species. Lesser degrees of homology can allow the construction of a phylogenetic tree, the extent to which any two species are related being reflected in their level of DNA homology. Cataloguing of nucleotide sequences of the highly conserved 16S and 23S ribosomal RNA molecules is also beginning to make an impact on taxonomic studies [2, 4] and these data on the relationships between nucleic acids of different species have provided a rational basis for the development of nucleic acid probes for identification purposes (see below).

The vast majority of published papers exploring the epidemiological relationship between mutans streptococci and caries have failed to pursue the characterization of isolates down to the level now required for precise allocation to species. However, from other studies on the frequency of isolation of different streptococci, it is clear that *S. mutans* is found in around 90% of people carrying mutans streptococci, while the frequency of carriage of *Streptococcus sobrinus* is considerably less – between 7 and 40% of isolates in various studies [1]. There is no evidence that any of the other species *S. cricetus, S. rattus, S. macacae, S. downei, S. ferus* play a significant role in caries in man. Nevertheless, there is a clear need for new simple and rapid procedures for counting and conclusively identifying mutans streptococci to further our knowledge of the contribution of different species to the aetiology of caries, investigate variations between population groups with different diets and different levels of diseases experience and to allow evaluation of bacteriological approaches to disease prognosis and detection of high-risk individuals. At the same time, there is now sufficient evidence to indicate that it is upon *S. mutans* that our more detailed attempts at biological characterization should be concentrated. *S. sobrinus* may play a subsidiary role and *S. sanguis* is also worthy of attention because it is such an important member of early dental plaques. This is not to say that all other species can be ignored – each in an individual way will take part in the myriad interactions taking place in plaque and so can ultimately have an influence on whether disease occurs or not, even though there is no direct link between them and caries.

*Periodontal Diseases*

The difficulties of assessing the significance of individual bacterial species isolated from subgingival plaque in the aetiology of periodontal diseases are also enormous, many isolates being strict anaerobes, nutritionally fastidious and hence hard to characterize and assign to appropriate taxonomic groups. Few laboratories have the expertise and resources to embark upon large-scale surveys but those that have, for example the groups led by Moore and by Socransky [5–8] have been able to show the regular occurrence of 300–400 distinct taxa. Analysis of the frequency of isolation from healthy and diseased sites in different patient groups indicates that perhaps only 10–20% of the species found may be disease-associated but the overall conclusion is that there are '... multiple possible aetiologies of periodontal destruction based on a select group of species that occur specifically or in greatly increased concentrations or frequency in diseased sites' [6]. Conversely, certain species seem to be indicators of gingival health [8]. Nevertheless, there is now sufficient accrued information to allow us to recognize some species as worthy of special attention, though the shifting emphasis on favourite species has led to the tag 'Bug of the Month Club' being applied to some enthusiasts! The potential virulence determinants of *Actinobacillus actinomycetemcomitans, Bacteroides gingivalis* and *Treponema denticola* have all attracted an increasing amount of attention in recent years though members of other genera which may be important, such as *Capnocytophaga, Eubacterium, Eikenella, Selenomonas, Wolinella* and *Peptostreptococcus* remain as yet virtually unexplored.

Because of the difficulties of culturing the strict anaerobes (and it is obvious that many new species still remain to be cultured and identified) and applying the appropriate tests needed for precise identification, there is a pressing need for new technologies which can be widely applied so that the skills of the microbiologist can be allied to those of the clinician to improve our ability to predict and monitor the progression (or recovery after treatment) of periodontal diseases.

### *Genetic Probes*

There is widespread interest in the potential usefulness of genetic probes in diagnosis of human diseases, both inherited ones and those due to infectious agents. The theoretical basis of such probes rests upon the nature of the DNA molecule, with its two complementary strands. The structure of DNA's double helix is determined by the strong bonding reaction between pairs of nucleotides in the opposing strands. If the two strands are separated, they can be reassociated under appropriate conditions either with each other or with any other fragment of DNA which has a complementary nucleotide sequence, to form a hybrid molecule. The possibility of availing of this phenomenon in diagnosis comes from the fact that the DNA of every living organism will possess unique sequences of nucleotides and so one can look to see whether a particular cell type contains a sequence which will hybridize with a 'probe' of known specificity. Furthermore, by altering the conditions under which hybridization is allowed to occur, it is possible to form hybrids between DNA sequences that are closely related though not identical. By such manipulations of the 'stringency' of reaction it is possible to classify a series of DNAs according to the closeness of their similarity.

DNA hybridization techniques have found wide application in molecular biology laboratories in the last decade, most commonly in the form of 'Southern blots' where DNA fragments are bound to nitrocellulose or nylon paper, treated to separate the strands and then hybridized with a DNA probe [9]. The probe is generally labelled by the incorporation of $^{32}P$ and the formation of hybrids can be recognized (after unbound probe is washed away) by autoradiography. Although solid-phase assays on nitrocellulose have found the widest application, alternative strategies for carrying out the hybridization in solution and then separating hybrids from probe have been developed and are already in use in some commercially available diagnostic kits [10]. Another area under active investigation is that of choice of label for the probe; high-energy radioisotopes can give great sensitivity but the hazards of working with them, their short half-life and high cost are serious disadvantages. Many alternative labelling systems, based on tagging the probe with enzymes, fluorochromes or haptenic groups which are subsequently recognized by antibody in a sandwich assay have been devised and are beginning to find wider application [10]. Such developments are essential if laboratories or dental practices lacking skills in molecular biology are to benefit from gene probe technology. In many ways, gene probes face the same type of problems as diagnostic probes based upon the specificity of monoclonal antibodies, and it seems likely that in at least some instances the latter may provide the basis of a preferable assay [11].

*Selection of DNA for a Probe*

For the generation of valid data it is essential that the specificity of a DNA probe be known. In the discussion of probe selection below, examples have been confined to the field of oral microbiology wherever possible.

*Whole Genomic Probes.* Studies of DNA/DNA hybridization in taxonomic studies are based upon reassociation of whole chromosomal (genomic) DNA under closely defined conditions. Although some hybridization between *S. mutans* and *S. sobrinus* DNA can be detected under such conditions, indicating a degree of relationship between the two species [12], it is possible to alter the stringency of the assay conditions so that labelled *S. mutans* DNA will bind only to DNA from strains of the same species. The more closely related two species are however, the more difficult it is to ensure the specificity of hybridization. In our laboratories we have been able to discriminate between *S. mutans, S. sobrinus* and other mutans streptococci using biotin-labelled genomic probes in a dot-blot assay [13]. For *Bacteroides* species, genomic probes also show a high level of specificity and in one study 94% of a collection of clinical isolates could be correctly identified [14]. A further advantage of the probe approach was that it could identify to species level several strains which could not be biochemically identified. In another report, it was found that genomic DNA from *B. intermedius* and *B. gingivalis* could be used as species-specific probes whereas *A. actinomycetemcomitans* lacked the requisite specificity and showed cross-hybridization with strains of *Haemophilus, Campylobacter* and *Wolinella* [15].

A major disadvantage of probes based on genomic DNA, apart from possible lack of specificity, is that it is necessary to routinely prepare DNA from the species on which the probe is based. This poses problems if the organism is difficult to culture or to extract DNA therefrom.

*Random Fragments of DNA.* Within the whole chromosome, certain genes or regions of DNA may be species-specific, while others are too closely related to those in other organisms to be useful as probes. One approach to finding the specific regions is therefore to cut the genome into small fragments (by sonication or by restriction endonucleases) and 'shotgun clone' the fragments into plasmid vectors. Individual clones from the resultant bank of random fragments are then screened to find ones which show hybridization of the desired specificity. This approach has been successfully applied [16] but can be further refined by the addition of a step analogous to absorption of sera in the preparation of reagents for serotyping, in which nonspecific DNA is removed by hybridizing to DNA from an organism related to the target one. For example, Schmidhuber et al. [17] allowed biotin-labelled DNA from *S. pneumoniae* to hybridize in solution with DNA fragments from *S. oralis.* Fragments which hybridized were removed by affinity binding to avidin-agarose and the nonhybridized fragments cloned into plasmids. Several of the plasmids obtained were then shown to be specific for *S. oralis*

*Specific Genes as Probes.* Many of the individual genes of oral bacteria which have been chosen for cloning (see below) code for surface antigens or enzymes which have already been the subject of comparisons between species. Others, once cloned, can easily be tested for their potential as probes. There may be many advantages in using known, characterized genes rather than random fragments of DNA, as they allow comparisons to be made between genetic and biochemical or serological means of typing isolates. It is, however, unlikely that direct comparison will be possible with most current serotyping schemes, since the latter are generally based upon polysaccharide antigens, the end-product of the activity of many genes.

*Ribosomal RNA.* Strands of DNA can form hybrids not only with DNA but also with RNA strands of complementary nucleotide sequence. Ribosomal RNA thus provides a potential target for probes and, because a growing cell may contain $10^4$ ribosomes, can provide much greater sensitivity compared to single-copy genes. With ribosome probes it is even possible to label and identify single cells [18]. Both 16S and 23S ribosomal RNA sequences have been highly conserved through evolution, though the subtle variations in sequence have provided the basis for most of our current ideas about lines of descent of bacterial species [4]. As the RNA sequences of target species become available, it is possible to identify the regions which are species-specific. Synthetic oligonucleotides (ca. 20 bases long) complementary to the specific region can then be constructed and used as probes [18, 19]. This may well turn out to be the method of choice for probe construction, once sequences of ribosomal RNA become available for more species.

### *Application of DNA Probes*

From the very promising results that have so far become available, it seems highly likely that DNA-probe based methods will soon begin to supplement or even supplant the lengthy series of biochemical tests currently needed for identification of many oral bacteria. Initially, the most straightforward application will be in the identification of strains which have been isolated in pure culture, as fewer problems of background cross-talk and low-level sensitivity will be experienced. Greatest progress has been made with the *Bacteroides*, but the basic knowledge for development of probes for other groups, *Streptococcus* in particular, is already available. For each new application, however, it will be essential to closely compare and validate data from DNA probes with conventional biochemical and serological tests.

Probes are also showing great potential as diagnostic agents by eliminating the need for culture of bacteria altogether. While yielding no information on the complexity of the microflora at a sampling site, probes can show the occurrence and level of particular organisms on which attention is focused. Thus, for example, as few as 100 cells of the suspect periodontopathogens *A. actinomycetemcomitans, B. intermedius* and *B. gingivalis* could be detected in single samples taken on papertip points from periodontal sites [20]. This level of detection is comparable to that which can be achieved with cultural methods [21]. If further work confirms the relevance of these particular species as indicator organisms, probe technology should thus allow earlier diagnosis of disease onset, indicate appropriate therapy and facilitate monitoring of the progress of treatment [22]. Nucleic acid probes may also be of value in diagnosis of oral viral infections, an application which has recently been reviewed [23].

### *DNA Fingerprinting*

The great capacity for variation in DNA caused by alterations in the sequence of its nucleotides means that, unless they are derived from the same clone, each isolate of a bacterium will have a unique DNA. This means that each isolate's DNA, when cut with restriction endonucleases, will give a characteristic set of fragments ('fingerprint') which can be separated by agarose gel electrophoresis. This approach of detecting restriction fragment length polymorphisms (RFLP) is finding wide application in human gene mapping and in all areas of population biology and has the potential for providing a powerful new tool for epidemiologists monitoring transmission of oral bacteria. In particular, it will make it possible to investigate the possibility that certain clonal lines of suggested pathogens are particularly associated with disease, as has been found for other bacteria [24]. Such DNA fingerprinting has already been applied to streptococci [25] and several laboratories are now investigating its value with oral species. Recent development of techniques for separating large fragments of DNA by field inversion gel electrophoresis [26] should lead to improved methods by providing simpler patterns for analysis.

## *Cloning of Genes from Oral Bacteria*

The uses to which microbiologists have put the new techniques of molecular biology are many and various, and no list of applications would ever be complete. However, amongst the advantages of cloning a gene (i.e. transferring it from one organism to another, incorporated into a vector molecule of some sort) are the following: (1) amplify production of the gene product; (2) analyze structure of the gene and its control sequences; (3) construct mutants with inactivated or altered genes, and (4) determine relationship to genes of other organisms.

The problems tackled by oral microbiologists have been no different in kind from those faced by workers with other groups of organisms, but each new species to be studied throws up its own particular set of challenges. Much ingenuity is required to make progress and it is often the case that the solution to a problem is best found from the literature in some quite different field of research. Starts have been made on the application of gene cloning to most of the main groups of oral bacteria, and there seems no reason why wider applications should not be seen in future.

### *Streptococci*

Ten years ago it was far from clear that streptococcal genes would be expressed effectively if introduced into that workhorse of molecular biology laboratories, *Escherichia coli*. This is one reason why efforts were put into studying the genetics of streptococci themselves, and the development of plasmid shuttle vectors which could replicate in either streptococci or *E. coli* [27, 28]. These vectors, and information on the best method to introduce foreign DNA into streptococci [29, 30], are proving of great benefit in strategies for gene inactivation (see below). Fundamental research into streptococcal genetics is not, however, the topic of this chapter.

The first demonstration that a gene from *S. mutans* could function well in *E. coli* came from Curtiss and co-workers [31] who showed that a plasmid-borne *asd* gene from *S. mutans* could provide aspartate semialdehyde dehydrogenase capable of carrying out the needed reaction in a mutant of *E. coli* lacking its own *asd* gene. The level of production of the *asd* gene product in recombinants was extremely high, representing 5–9% of the total *E. coli* protein [32], indicating that *S. mutans* control signals – promoters and ribosome-binding sites – were recognized by the *E. coli* transcription and translation machinery. This was later confirmed by DNA-sequencing and mRNA-transcript studies [33] and indeed the *asd* promoter region is so efficient that it could be used to obtain expression of *Mycobacterium leprae* genes in *E. coli* [34]. Several other genes, including *gtfI* from *S. downei* (fig. 1), have also been shown to be expressed at high level under their own control in *E. coli*. Not enough studies have yet been carried out, however, to allow the identification of consensus promoter sequences in streptococci or indeed even to identify the precise sites at which transcription starts.

Because of the significance of sucrose in the causation of caries, and the importance of interactions between oral streptococci and their host, the main areas of gene-cloning activity with oral streptococci have been in analyzing sucrose metabolism and characterizing surface components.

*Sucrose Metabolism.* It is perhaps fortunate for us that *E. coli* K-12, the strain most commonly used for genetic work, is unable to utilize sucrose. However, since sucrose is not transported across the *E. coli* cell membrane, indirect methods must be devised to screen for recombinants in which streptococcal sucrases have been expressed from plasmid-borne genes. First, one can seek recombinants which break down sucrose in the periplasm – this would be expected to work only if the streptococcal enzyme had been secreted (or leaked) through the cytoplasmic membrane and was successfully used by Burne et al. [35, 36]. Secondly, one can use the scheme devised by Robeson et al. [37] in which the trisaccharide raffinose (galactose-glucose-fructose) is taken up by the cell and the galactose cleaved off by the *E. coli*'s α-galactosidase effectively leaving sucrose. A *gal* mutation prevents the galactose being used as carbon source, so any recombinant colonies which show fermentative ability on raffinose must be doing so because they are expressing a streptococcal sucrase. Thirdly, it is possible to release any sucrose-

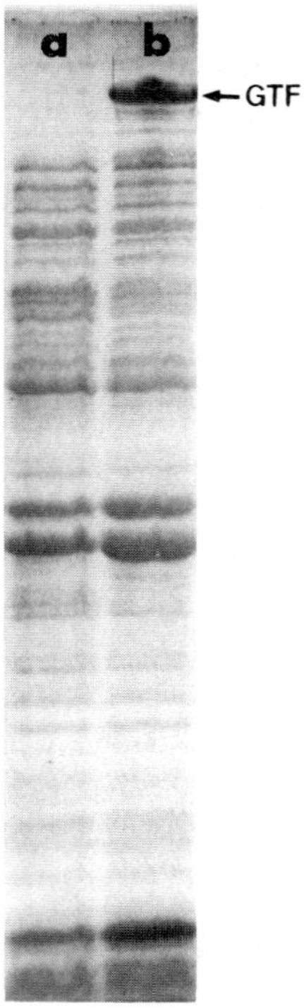

1

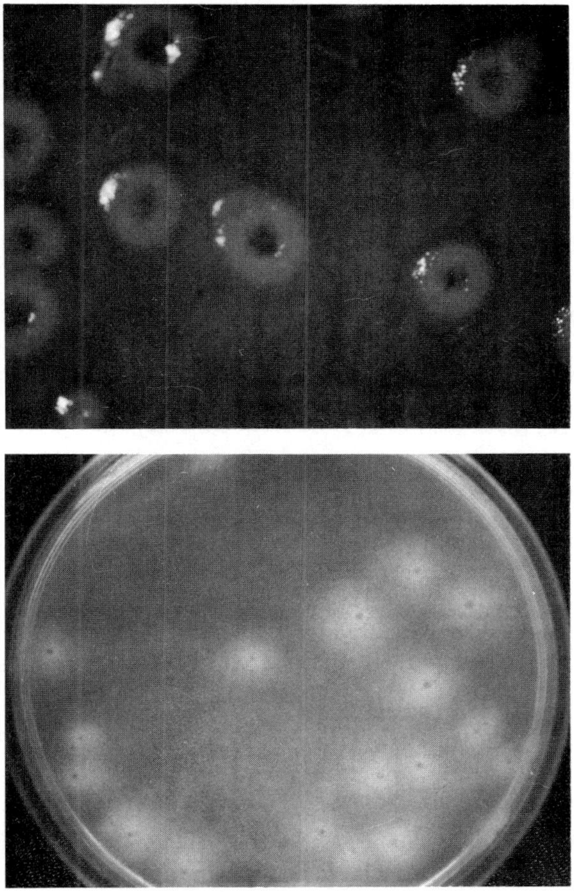
2a
2b

*Fig. 1.* SDS-polyacrylamide gel electrophoresis of protein from *(a) E. coli* and *(b) E. coli* carrying a plasmid carrying the *gtfI* gene for glucosyltransferase from *S. downei* [45] showing the efficient expression of the streptococcal gene.

*Fig. 2.* Appearance of plaques formed on *E. coli* grown on sucrose-minimal medium and infected with recombinant lambda bacteriophage: *(a)* cross-feeding due to breakdown of sucrose by the product of the *S. mutans gtfA* sucrose phosphorylase gene [38, 62]; *(b)* accumulation of polymer around recombinant plaques in which glucosyltransferase is being produced [40, 41].

metabolizing enzyme from *E. coli* by expressing the cloned gene from a recombinant bacteriophage or cosmid vector [38, 39]. Lysis of infected cells releases the enzyme which can then attack sucrose in the growth medium. Generation of glucose and/or fructose can then be detected either by using tetrazolium to detect reducing sugars [40], accumulation of polymer [41], or by cross-feeding of the surrounding uninfected *E. coli* [38] (fig. 2). A fourth screening method, applicable only for the detection of proteins which have already been purified, is to use specific antibody in an immunodetection method [41, 42]. Each of these screening approaches may show a bias towards the detection of a particular class of recombinant gene and it is noteworthy that only the 'phage method has yielded recombinants expressing genes for the polymer-forming glucosyltransferases (GTF) and fructosyltransferases, even though it was subsequently shown that these could also be expressed from plasmid vectors.

Major questions about the important group of extracellular GTF, which conventional biochemical approaches have been unable to answer, include the number of different GTF in each species of streptococcus and the basis for the formation of glucans with different proportions of 1,3- and 1,6-linkages [43]. Two distinct *gtf* genes have so far been cloned from *S. downei* MFe28 (formerly classified as *S. sobrinus*) [41, 44–46] and two from *S. mutans* GS-5 [47–49] but in both organisms there is evidence for the existence of further *gtf* genes. *S. salivarius* also has several distinct *gtf* genes [Jacques et al., submitted] but it appears that *S. sanguis* has only one [50].

When studying GTF produced by streptococci, it is well-nigh impossible to determine the activity of individual enzymes in crude preparations when several GTF may contribute to synthesis of a complex glucan. Once individual *gtf* genes have been cloned, however, recombinant *E. coli* can be used to produce unlimited quantities of a single GTF which can be readily purified for detailed characterization [45]. Comparisons can also be made between different purified enzymes and, at a more fundamental level, the nucleotide sequence of different *gtf* genes can be compared. It has been found, for example, that the genes *gtfI* from *S. downei* and *gtfB* from *S. mutans* show nearly 60% base-sequence homology [51]. This indicates a close evolutionary relationship but also suggests that the nonmatching regions will be those coding for the characteristic features of each gene product while the common stretches will determine the properties found in both GTFs. As further sequences become available, many long-standing problems about this interesting group of enzymes should be resolved.

Besides the GTF, other extracellular *S. mutans* enzymes for which the genes have been cloned are fructosyltransferase [52, 53] and a fructanase [54], while from *S. sobrinus* a dextranase gene has been cloned [55].

There have as yet not been any reports of the cloning of genes for sucrose transport systems, though both our laboratory and Kuramitsu [pers. commun.] have evidence for their existence closely linked to genes encoding enzymes for the further metabolism of sucrose within the cell. Sucrose can, of course, be used as a carbon source after being broken down to glucose and fructose outside the cell but two routes for its direct utilization are now also known. In one, sucrose is phosphorylated during uptake by the PTS transport system to give sucrose-6-phosphate inside the cell. This is then metabolized to glucose-6-phosphate and fructose by the product of gene *scrB*, which has been cloned by several groups [38, 56, 57] and recently sequenced [58]. It will be interesting to learn whether this gene is linked to others needed for the same pathway, as has been found in *E. coli* and *Klebsiella* [59, 60, 61]. The second sucrose pathway is via a non-PTS transport system which brings sucrose across the membrane. Sucrose then reacts with inorganic phosphate to form glucose-1-phosphate and fructose, a reaction catalyzed by sucrose phosphorylase (EC 2.4.1.7) [62]. This enzyme is encoded by gene *gtfA*, which was long thought to determine a glucan-forming enzyme [37, 38, 63]. The confusion arose because glucose-1-phosphate is insoluble in organic solvents and so behaves in the same way as glucans in the usual assay for detecting products made from $^{14}$C-sucrose. Adjacent to *gtfA*, and probably cotranscribed with it [64], is a gene which we believe to be involved in membrane transport because it shows a high degree of homology to an *E. coli* transport gene. Also closely linked is a dextranase gene [35] which we have found to encode an intracellular exodextranase. This releases glucose units from the reducing termini of dextran or (the more likely natural substrates) isomalto-saccharides of 2–7 units which have been transported across the membrane.

There is thus an accumulating amount of evidence for the close linkage of genes with related functions, and possibly operon-like arrangements of genes which are under coordinate regulation. Within the next few years it can be anticipated that much more information from gene sequencing will become available, which will help us to identify the unique features of gene control in streptococci as well as the similarities to other genera, and throw light upon the modulation of the multiplicity of catabolic and synthetic pathways for which sucrose is the starting point. Another interesting topic to be explored is the evolutionary origin of the genes involved in sucrose metabolism and the possibility that the oral streptococci acquired them relatively recently, along with their ability to colonize the mouth and utilize the substrates there. Comparisons with free-living species of *Bacillus* and *Leuconostoc*, which can also metabolize sucrose, should be informative here. The currently known routes of sucrose metabolism, and the steps for which genes have been cloned, are shown in figure 3.

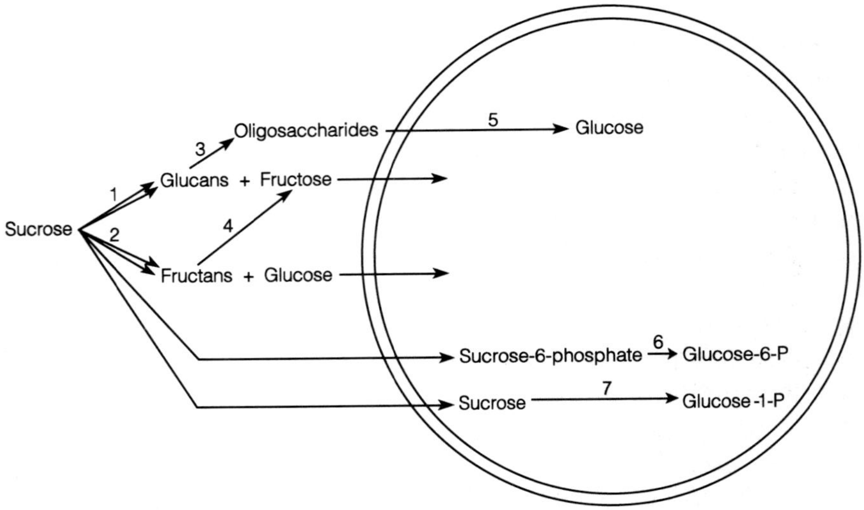


*Fig. 3.* Pathways for utilization of sucrose by *S. mutans* showing the steps for which genes have been cloned: 1 = glucosyltransferases, *gtfB, gtfC* [47–49]; 2 = fructosyltransferases, *ftf, gbp* [42, 52, 53]; 3 = dextranase, *dexA*; 4 = fructanase, *fruA* [54]; 5 = dextran glucosidase, *dexB* [35]; 6 = sucrose-6-phosphate hydrolase, *scrB* [56, 57]; 7 = sucrose phosphorylase, *gtfA* [62].

*Surface Proteins.* Surface protein antigens of oral streptococci have attracted research interest because of their involvement in specific interactions in adherence processes in dental plaque and their possible value in dental caries vaccines [65–67]. The gene for an antigen of 190,000 MW, *spaA*, has been cloned from *S. sobrinus* [68, 69] into *E. coli*. This gene has also been inserted into an avirulent strain of *Salmonella typhimurium* for experiments in the use of such a recombinant as an oral vaccine. The gene for the corresponding cross-reactive high molecular weight antigen has also been cloned from *S. mutans* [40, 70]. The *S. mutans* gene for a saliva-receptor protein, which may possibly prove to be the same as the 190,000 MW antigen [Russell and Klein, unpubl. data], has also been cloned [71]. Another wall-associated protein gene, that for antigen A [72], has recently been sequenced by my laboratory in collaboration with Ferretti and My Lien Dao. In addition to providing information on the structure of this protein and its probable means of interacting with the membrane and peptidoglycan layers, cloning of the gene has allowed us to isolate the full-size native antigen from recombinant *E. coli*, something that has been impossible in *S. mutans* because of degradation and complexing with other wall components.

*Gene Inactivation Studies.* Much can be learnt from biochemical investigation of streptococcal products expressed singly in *E. coli*, but for full understanding of their importance in overall cellular physiology it is advantageous to construct mutants lacking the specific function. Naturally occurring and chemically induced mutants have already been widely employed in the study of oral streptococci [32, 73] but genetic engineering methods can permit considerable improvement by allowing the targeted inactivation of known genes and giving confidence that multiple mutations have not occurred. Methods are now in use for inactivating genes cloned in *E. coli* by insertion of extraneous DNA [56, 74], or by deletion of a portion of a gene [63, 75]. When such modified genes are reintroduced to the original streptococcus by transformation with a shuttle vector which cannot replicate, forced integration and recombination results in interruption of the chromosomal gene and hence loss of its function. Insertion of transposons (mobile genetic elements, such as Tn916 which confers resistance to tetracycline) have also been used to inactivate target genes [50, 76]. In addition to their value for fundamental studies of bacterial physiology, genetically engineered mutants lacking specific virulence determinants or with otherwise altered properties may find an application in strategies of replacement therapy where an avirulent mutant is used to displace pathogens [77].

*Other Oral Bacteria*

Although the greatest amount of effort has so far been devoted to studies of oral streptococci, the feasibility of applying similar techniques to other groups of bacteria has been demonstrated by a few reports and it is clear that work on *Lactobacillus* and *Actinomyces* will benefit from a genetic approach [78, 79]. The anaerobes can be anticipated to provide a distinct set of challenges, but research, particularly on transmissible antibiotic resistance in gut *Bacteroides*, has laid the basis of an understanding [80] and a report of the cloning of a fimbrial subunit from *Bacteroides gingivalis* has recently appeared [81]. Several other laboratories are also known to have successfully cloned genes from *Bacteroides*. Finally, oral spirochaetes of the genus *Treponema*, a group in which there is virtually no information on genetics, is also yielding to a gene-cloning approach and we have expressed both the trypsin-like enzyme and axial filament peptides in *E. coli* [82].

As in other areas of biology, techniques for the manipulation of DNA are rapidly proving their worth in many aspects of oral microbiology. The last few years have seen a rapid rise in the number of publications describing the cloning of genes and much additional information of value is now being acquired from nucleotide-sequencing, gene-inactivation and site-directed mutagenesis studies. The range of organisms being examined is steadily widening and we can look forward to many exciting discoveries and beneficial applications in the years ahead.

## *References*


1 Loesche WJ: Role of *Streptococcus mutans* in human dental decay. Microbiol Revs 1986;50:353–380.

2 Schleifer KH, Kilpper-Balz R: Molecular and chemotaxonomic approaches to the classification of streptococci, enterococci and lactococci: a review. Syst Appl Microbiol 1987;10:1–19.

3 Coykendall AL: Genetic heterogeneity in *Streptococcus mutans*. J Bacteriol 1971; 106:192–196.

4 Woese CR: Bacterial evolution. Microbiol Rev 1987;51:221–271.

5 Moore WEC: Microbiology of periodontal disease. J Periodont Res 1987;22:335–341.

6 Moore WEC, Holdeman LV, Cato EP, Smibert RM, Burmeister JA, Palcanis KG, Ranney RR: Comparative bacteriology of juvenile periodontitis. Infect Immun 1985; 48:507–519.

7 Sockransky SS, Haffajee AD, Smith GLF, Dzink JL: Difficulties encountered in the search for the etiologic agents of destructive periodontal diseases. J Clin Periodontol 1987;14:588–593.

8 Dzink JL, Socransky SS, Haffajee AD: The predominant cultivable microbiota of active and inactive lesions of destructive periodontal diseases. J Clin Periodontol 1988;15:316–323.

9 Hames BD, Higgins SJ: Nucleic Acid Hybridisation: A Practical Approach. Oxford, IRL Press, 1985.

10 Matthews JA, Kricka LJ: Analytical strategies for the use of DNA probes. Anal Biochem 1988;169:1–25.

11 Edberg SC: Principles of nucleic acid hybridization and comparison with monoclonal antibody technology for the diagnosis of infectious diseases. Yale J Biol Med 1985;58:425–442.

12 Coykendall AL: Four types of *Streptococcus mutans* based on their genetic, antigenic, and biochemical characteristics. J Gen Microbiol 1974;83:327–338.

13 Ivic A, Russell RRB: The use of DNA probes for the identification of oral microorganisms. J Dent Res 1988;67:670.

14 Roberts MC, Moncla B, Kenny GE: Chromosomal DNA probes for the identification of *Bacteroides* species. J Gen Microbiol 1987;133:1423–1430.

15 Strzempko MN, Simon SL, French CK, Lippke JA, Raia FF, Savitt ED, Vaccaro KK: A cross-reactivity study of whole genomic DNA probes for *Haemophilus actinomy-*

*cetemcomitans*, *Bacteroides intermedius* and *Bacteroides gingivalis*. J Dent Res 1987;66:1543–1546.

16 Kuritza AP, Getty CE, Shaughnessy P, Hesse R, Salyers AA: DNA probes for identification of clinically important *Bacteroides* species. J Clin Microbiol 1986;23:343–349.

17 Schmidhuber S, Ludwig W, Schleifer KH: Construction of a DNA probe for the specific identification of *Streptococcus oralis*. J Clin Microbiol 1988;26:1042–1044.

18 Giovannoni SJ, DeLong EF, Olsen GJ, Pace NR: Phylogenetic group-specific oligodeoxynucleotide probes for identification of single microbial cells. J Bacteriol 1988;170:720–726.

19 Regensburger A, Ludwig W, Schleifer KH: DNA probes with different specificities from a cloned 23S rRNA gene of *Micrococcus luteus*. J Gen Microbiol 1988;134:1197–1204.

20 French CK, Savitt ED, Simon SL, Eklund SM, Chen MC, Klotz LC, Vaccaro KK: DNA probe detection of periodontal pathogens. Oral Microbiol Immunol 1986;1:58–62.

21 Savitt EK, Strzempko MN, Vaccaro KK, Peros WJ, French CK: Comparison of cultural methods and DNA probe analyses for the detection of *Actinobacillus actinomycetemcomitans*, *Bacteroides gingivalis* and *Bacteroides intermedius* in subgingival plaque samples. J Periodontol 1988;59:431–438.

22 Peros WJ, Savitt ED, Vassos G, Milligan R, Niederman R: Rapid microbiologic tests as an adjunct to the diagnosis of periodontal disease. Compend Cont Educ Dent 1988;9:234–241.

23 Maitland NJ, Cox MF, Lynas C, Prime S, Crane I, Scully C: Nucleic acid probes in the study of latent viral disease. J Oral Pathol 1987;16:199–211.

24 Selander RK, Musser JM, Caugant DA, Gilmour MN, Whittam TS: Population genetics of pathogenic bacteria. Microb Pathol 1987;3:1–7.

25 Skjold SA, Quie PG, Fries LA, Barnham M, Cleary PP: DNA fingerprinting of *Streptococcus zooepidemicus* (Lancefield group C) as an aid to epidemiological study. J Infect Dis 1987;155:1145–1150.

26 Grothues D, Tummler B: Genome analysis of *Pseudomonas aeruginosa* by field inversion gel electrophoresis. FEMS Microbiol Lett 1987;48:419–422.

27 Macrina FL, Evans RP, Tobian JA, Hartley DL, Clewell DB, Jones KR: Novel shuttle plasmid vehicles for *Escherichia-Streptococcus* transgeneric cloning. Gene 1986;25:145–150.

28 Dao ML, Ferretti JJ: *Streptococcus-Escherichia coli* shuttle vector pSA3 and its use in the cloning of streptococcal genes. Appl Environ Microbiol 1985;49:115–119.

29 Murchison HH, Barrett JF, Cardineau GA, Curtiss R: Transformation of *Streptococcus mutans* with chromosomal and shuttle plasmid (pYA629) DNAs. Infect Immun 1986;54:273–282.

30 Lindler LE, Macrina FL: Characterization of genetic transformation in *Streptococcus mutans* by using a novel high-efficiency plasmid marker rescue system. J Bacteriol 1986;166:658–665.

31 Jaqusztyn-Krynicka EK, Smorawinska M, Curtiss R: Expression of *Streptococcus mutans* aspartate-semialdehyde dehydrogenase cloned into plasmid pBR322. J Gen Microbiol 1982;128:1135–1145.

32 Curtiss R: Genetic analysis of *Streptococcus mutans* virulence; in Goebel W (ed): Genetic Approaches to Microbial Pathogenicity. Curr Top Microbiol Immunol. Berlin, Springer, 1985, vol 118, pp 253–277.

33 Cardineau GA, Curtiss R: Nucleotide sequence of the *asd* gene of *Streptococcus mutans.* J Biol Chem 1987;262:3344–3353.
34 Jacobs WR, Docherty MA, Curtiss R, Clark-Curtiss JE: Expression of *Mycobacterium leprae* genes from a *Streptococcus mutans* promoter in *Escherichia coli* K-12. Proc Natl Acad Sci USA 1986;83:1926–1930.
35 Burne RA, Rubinfield B, Bowen WH, Yasbin RE: Tight genetic linkage of a glucosyltransferase and dextranase of *Streptococcus mutans* GS-5. J Dent Res 1986; 65:1392–1401.
36 Burne RA, Schilling K, Bowen WH, Yasbin RE: Expression, purification, and characterization of an exo-β-*D*-fructosidase of *Streptococcus mutans.* Infect Immun 1987; 169:4507–4517.
37 Robeson JP, Barletta RG, Curtiss R: Expression of a *Streptococcus mutans* glycosyltransferase gene in *Escherichia coli.* J Bacteriol 1983;153:211–221.
38 Russell RRB, Morrissey P, Dougan G: Cloning of sucrase genes from *Streptococcus mutans* in bacteriophage lambda. FEMS Microbiol Lett 1985;30:37–41.
39 Jacobs WR, Barrett JF, Clark-Curtiss JE, Curtiss R: In vivo repackaging of recombinant cosmid molecules for analyses of *Salmonella typhimurium, Streptococcus mutans,* and mycobacterial gene libraries. Infect Immun 1986;52:101–109.
40 Morrissey P, Dougan R, Russell RRB, Gilpin M: Cloning of *Streptococcus mutans* antigens in *Escherichia coli* as an aid to the development of a dental caries vaccine; in Lerner RA, Chanock RM, Brown F (eds): Vaccines 85. New York, Cold Spring Harbor Laboratories, 1986, pp 117–120.
41 Gilpin M, Morrissey P, Russell RRB: Cloning and expression of two *Streptococcus mutans* glucosyltransferases in *Escherichia coli* K-12. Infect Immun 1985;49:414–416.
42 Russell RRB, Coleman D, Dougan G: Expression of a gene for glucan-binding protein from *Streptococcus mutans* in *Escherichia coli.* J Gen Microbiol 1985;131: 295–299.
43 Mukasa H: Properties of *Streptococcus mutans* glycosyltransferases; in Hamada S, Michalek S, Kiyono H, Menaker L, McGhee JR (eds): Molecular Microbiology and Immunobiology of *Streptococcus mutans.* New York, Elsevier, 1986, pp 121–132.
44 Whiley RA, Russell RRB, Hardie JM, Beighton D: *Streptococcus downei* sp.nov. of strains previously described as *Streptococcus mutans* serotype *h.* Int J Syst Bacteriol 1988;38:25–29.
45 Russell RRB, Gilpin ML, Mukasa H, Dougan G: Characterization of glucosyltransferase expressed from a *Streptococcus sobrinus* gene cloned in *Escherichia coli.* J Gen Microbiol 1987;133:935–944.
46 Ferretti JJ, Gilpin ML, Russell RRB: Nucleotide sequence of a glucosyltransferase gene from *Streptococcus sobrinus* MFe28. J Bacteriol 1987;169:4271–4278.
47 Aoki H, Shiroza T, Hayakawa M, Sato S, Kuramitsu HK: Cloning of a *Streptococcus mutans* glucosyltransferase gene coding for insoluble glucan synthesis. Infect Immun 1986;53:587–594.
48 Pucci MJ, Jones KR, Kuramitsu HK, Macrina FL: Molecular cloning and characterization of the glucosyltransferase C gene *(gtfC)* from *Streptococcus mutans* LM7. Infect Immun 1987;55:2176–2182.
49 Hanada N, Kuramitsu HK: Isolation and characterization of the *Streptococcus mutans gtfC* gene coding for both soluble and insoluble glucan synthetic activity. Infect Immun 1988;56:1999–2005.

50 Ivic A, Smith K, Russell RRB: Transposon mutagenesis of *S. sanguis* glucosyltransferase. Xth Lancefield International Symposium on Streptococci and Streptococcal Diseases, Köln, 1987 (in press).

51 Russell RRB, Shiroza T, Kuramitsu HK, Ferretti JJ: Homology of glucosyltransferase gene and protein sequences from *Streptococcus sobrinus* and *Streptococcus mutans*. J Dent Res 1988;67:543–547.

52 Shiroza T, Kuramitsu HK: Sequence analysis of the *Streptococcus mutans* fructosyltransferase gene and flanking regions. J Bacteriol 1988;170:810–816.

53 Russell RRB, Gilpin ML: Identification of virulence components of mutans streptococci; in Ferretti JJ, Curtiss R (eds): Streptococcal Genetics. Washington, American Society for Microbiology, 1987, pp 201–204.

54 Burne RA, Schilling K, Bowen WH, Yasbin RE: Expression, purification, and characterization of an exo-β-*D*-fructosidase of *Streptococcus mutans*. J Bacteriol 1987;169:4507–4517.

55 Barrett JF, Barrett TA, Curtiss R: Purification and partial characterization of the multicomponent dextranase complex of *Streptococcus sobrinus* and cloning of the dextranase gene. Infect Immun 1987;55:792–802.

56 Lunsford RD, Macrina FL: Molecular cloning and characterization of *scrB*, the structural gene for the *Streptococcus mutans* phosphoenolpyruvate-dependent sucrose phosphotransferase system sucrose-6-phosphate hydrolase. J Bacteriol 1986;166:426–434.

57 Hayakawa M, Aoki H, Kuramitsu HK: Isolation and characterization of the sucrose-6-phosphate hydrolase gene from *Streptococcus mutans*. Infect Immun 1986;53:582–586.

58 Sato Y, Kuramitsu HK: Sequence analysis of the *Streptococcus mutans scrB* gene. Infect Immun 1988;56:1956–1960.

59 Schmid K, Ebner R, Altenbuchner J, Schmitt R, Lengeler JW: Plasmid-mediated sucrose metabolism in *Escherichia coli* K12: mapping of the *scr* genes of pUR400. Mol Microbiol 1988;2:1–8.

60 Hardesty C, Colon G, Ferran C, DiRienzo JM: Deletion analysis of sucrose metabolic genes from a *Salmonella* plasmid cloned in *Escherichia coli* K12. Plasmid 1987;18:142–155.

61 Sprenger GA, Lengeler JW: Analysis of sucrose catabolism in *Klebsiella pneumoniae* and in $Scr^+$ derivatives of *Escherichia coli* K12. J Gen Microbiol 1988;134:1635–1644.

62 Russell RRB, Mukasa H, Shimamura A, Ferretti JJ: *Streptococcus mutans gtfA* gene specifies sucrose phosphorylase. Infect Immun 1988;56:2763–2765.

63 Pucci MJ, Macrina FL: Molecular organization and expression of the *gtfA* gene of *Streptococcus mutans* LM7. Infect Immun 1986;54:77–84.

64 Ferretti JJ, Huang T-T, Russell RRB: Sequence analysis of the glucosyltransferase A *(gtfA)* gene from *Streptococcus mutans* Ingbritt. Infect Immun 1988;56:1585–1588.

65 Russell RRB, Abdulla A, Gilpin ML, Smith K: Characterization of *Streptococcus mutans* surface antigens; in Hamada S, Michalek S, Kiyono H, Menaker L, McGhee JR (eds): Molecular Microbiology and Immunobiology of *Streptococcus mutans*. New York, Elsevier, 1986, pp 61–70.

66 Krasse B, Emilson C-G, Gahnberg L: An anticaries vaccine: report on the status of research. Caries Res 1987;21:255–276.

67 Russell RRB, Johnson NW: The prospects for vaccination against dental caries. Br Dent J 1987;162:29–34.
68 Holt RG, Abiko Y, Saito S, Smorawinska M, Hansen JB, Curtiss R: *Streptococcus mutans* genes that code for extracellular proteins in *Escherichia coli* K-12. Infect Immun 1982;38:147–156.
69 Holt RG, Ogundipe JO: Molecular cloning in *Escherichia coli* of the gene for a *Streptococcus sobrinus* surface protein containing two antigenic determinants; in Ferretti JJ, Curtiss R (eds): Streptococcal Genetics. Washington, American Society for Microbiology, 1987, pp 217–219.
70 Lee FS, Progulske-Fox A, Bleiweis AS: Molecular cloning and expression of a major surface protein antigen P1 from *Streptococcus mutans* in *Escherichia coli.* Infect Immun 1988;56:2114–2119.
71 Sommer P, Bruyere T, Ogier JA, Garnier J-M, Jeltsch J-M, Klein J-P: Cloning of the saliva-interacting protein gene from *Streptococcus mutans.* J Bacteriol 1987;169: 5167–5173.
72 Russell RRB: Wall-associated protein antigens of *Streptococcus mutans.* J Gen Microbiol 1979;114:109–115.
73 Freedman ML, Tanzer JM, Coykendall AL: The use of genetic variants in the study of dental caries; in Tanzer JM (ed): Animal Models in Cariology. Washington DC, Information Retrieval, 1981, pp 247–269.
74 Kuramitsu HK: Utilization of a mini-mu transposon to construct defined mutants in *Streptococcus mutans.* Mol Microbiol 1987;1:229–231.
75 Sato S, Ueda S, Kuramitsu HK: Construction of *Streptococcus mutans* glucosyltransferase mutants utilizing a cloned gene fragment. FEMS Microbiol Lett 1987;48:207 210.
76 Gawron-Burke C, Clewell DB: Regeneration of insertionally-inactivated streptococcal DNA fragments after excision of transposon Tn*916* in *Escherichia coli*: strategy for targeting and cloning of genes from gram-positive bacteria. J Bacteriol 1984;159: 214–221.
77 Hillman JD, Sockransky SS: Replacement therapy for the prevention of dental disease. Adv Dent Res 1987;1:119–125.
78 Chassy BM, Flickinger JL: Transformation of *Lactobacillus casei* by electroporation. FEMS Microbiol Lett 1987;44:173–177.
79 Donkersloot JA, Cisar JO, Wax ME, Harr RJ, Chassy BM: Expression of *Actinomyces viscosus* antigens in *Escherichia coli*: cloning of a structural gene *(fimA)* for type 2 fimbriae. J Bacteriol 1985;162:1075–1078.
80 Odelson DA, Rasmussen JL, Smith CJ, Macrina FL: Extrachromosomal systems and gene transmission in anaerobic bacteria. Plasmid 1987;17:87–109.
81 Dickinson DP, Kubiniec MA, Yoshimura F, Genco RJ: Molecular cloning and sequencing of the gene encoding the fimbrial subunit protein of *Bacteroides gingivalis.* J Bacteriol 1988;170:1658–1665.
82 MacDougall JA, Russell RRB: Trypsin-like enzyme from oral *Treponema denticola.* J Dent Res 1988;67:654.


R.R.B. Russell, PhD, Hunterian Dental Research Unit, London Hospital Medical College, Turner Street, GB-London E1 2AD (UK)

Ferguson DB (ed): Aspects of Oral Molecular Biology.
Front Oral Physiol. Basel, Karger, 1991, vol 8, pp 77–94

# Genetic Control of Salivary Mucin Formation[1]

*L.A. Tabak*

Departments of Dental Research and Biochemistry,
University of Rochester, N.Y., USA

Epithelial surfaces are protected by a slimy, viscoelastic coat termed *mucus*. This adherent layer, consisting of glycoproteins, proteins and lipids, forms a jelly-like blanket that serves as the first line of defense against mechanical, chemical, or microbial insult to the underlying tissues [1]. The unique physicochemical and rheological properties of the mucus coat are attributed largely to the high molecular weight, heavily glycosylated *mucin-glycoproteins* (mucins).

Within the mouth, mucins lubricate and hydrate the hard and soft tissues, providing a demulcent coating which fends off a wide range of exogenous insults [2, 3]. The mucosal dryness and sensation of 'burning', as well as the difficulty in phonation, food bolus formation and deglutition which accompany human salivary hypofunction, have been linked to a quantitative lack of, or qualitative change in, salivary mucins [4, 5].

Although considerable progress has been made in delineating the structure (for example, human [6, 7], monkey [8, 9], mouse [10, 11], ovine [12, 13], porcine [14, 15], rat [15, 17]) and conformation [18–24] of salivary mucins as well as events surrounding mucin secretion [for review, see 25], details relating to the control of salivary mucin expression are only beginning to emerge.

This chapter reviews the current knowledge of the mechanisms which underlie salivary mucin biosynthesis. While recent molecular and biochemical studies have provided a much needed conceptual framework, it is clear

[1] Original work cited herein was supported by USPHS Grants DE 08108, DE 08511, and Research Career Development Award DE 00217. I wish to acknowledge Drs. W.H. Bowen, L. Cooper, D. Culp, and B. O'Connell for their careful review of the manuscript. I thank Ms. Beth Tollers for typing the manuscript.

that additional information will be required before the available data can be placed into a coherent physiological context. Therefore, we will attempt to point out those gaps in our current understanding that appear to merit most attention for future study.

### *The Unique Composition and Structure of Mucins Presents a Regulatory Challenge in Biosynthetic Control*

Mucins are high molecular weight glycoproteins which share common features including a peptide core (apomucin) enriched in serine, threonine and proline residues. The apomucin undergoes a wide variety of co- and posttranslational modifications including acylation, glycosylation and sulfation. Carbohydrate side chains (oligosaccharides) which are linked O-glycosidically to threonine and/or serine may contain galactose (Gal), fucose (Fuc), N-acetylgalactosamine (GalNac), N-acetylglucosamine (GlcNac), and sialic acids (SA). Sulfate, if present, is commonly linked to either Gal or GlcNac. In addition, the presence of a small number of asparagine-linked (N-linked) oligosaccharides has been reported [10, 11]. Once synthesized, mucin subunits enter into both heterotypic as well as homotypic interactions which result in the formation of elaborate suprastructures. The molecular suprastructure, which may be stabilized by either (or both) noncovalent or covalent forces, is responsible for the unique rheological properties (i.e. elasticity, adhesiveness, insolubility) which are characteristic of mucin-containing solutions [1, 2]. The extensive array of co- and posttranslational modifications, as well as the process of mucin assembly, places severe limitation into the regulatory mechanisms which underlie production. In contrast to proteins and nucleic acids which are synthesized by highly accurate template mechanisms, much of the genetic information required for the expression of mucins is transferred via *indirect* pathways [26]. Thus, in addition to considering the events which regulate the gene(s) which code for the apomucin, one must also take into account the genes encoding those enzymes involved in the co- and posttranslational modifications such as the glycosyltransferases (and in the case of N-linked oligosaccharides, the glycosidases), the acyltransferases and the sulfonyltransferases. These multiple points of control exerted at the gene level are superimposed over additional potential points of regulation, including availability of reactants for the modifying enzymes (e.g. glycosyltransferase). Collectively, the existence of multiple control points (perhaps not tightly or coordinately regulated) results in a certain degree of variation (termed microheterogeneity) in the assembly of mucins. How contrived this heterogeneity really is has been debated [2, 27, 28]; until a specific function can be attributed to each specific carbohydrate variant, however, the arguments remain speculative.

### *Biosynthesis of Apomucin*

#### *Composition and Structure of Apomucin*

The very properties which endow mucins with protective qualities render them largely intractable to conventional chemical analysis. In particular, the propensity of mucins to complex heterotypically [for examples, see 6, 8, 12, 16] makes it essential to evaluate the homogeneity of a mucin preparation by multiple, complementary criteria [8, 12, 16]. Often, there are significant amounts of contaminating peptides which are removed only after the suprastructure of the mucin architecture is altered; presumably these peptides are entrapped in the mucin meshwork and they are released by such treatments as reduction of disulfide bands [8], derivitization of ε-amino groups of lysine (with dansyl HCl [12] or reductive methylation [6, 16]), or deglycosylation of the mucin [15].

Amino acid analysis of highly purified salivary mucins reveals high levels of Ser, Thr, and Pro with very low levels (or the absence of) aromatic amino acids, Met and His. In general, the more highly purified the mucin, the more degenerate the number of different amino acids observed [15, 16]. Only limited sequence analysis is available for salivary mucins; nevertheless, some common patterns have emerged. In both bovine [29] and porcine [15, 30] submandibular gland mucin the protein core has been shown to be made up of a repeating unit of amino acids. The presence of such repeating units is characteristic of many salivary protein families, e.g. proline-rich proteins [31], the Drosophila glue proteins [32], SPT proteins [33], and glutamine-glutamic acid-rich proteins [34, 35]. All of these proteins are 'soluble structural proteins', i.e. they appear to serve a tissue-coating function within the oral cavity. The events leading to the evolution of a tandem repeat motif have been the subject of speculation [30, 33]; in general, it is thought that a given array of amino acids can perform the biological function of these molecules and with no negative evolutionary pressure, the number of repeats increases. Thus, in the case of mucins, if the protein core is mostly serving a 'scaffolding' function, for the attachment of oligosaccharides, additional segments would not be harmful and, indeed, may be advantageous.

While one report has appeared describing the 'globular' nature of deglycosylated ovine submandibular gland mucin [19], more recent work has suggested that the peptide backbone is best described as an extended form of a random coil. For example, the apomucin of the porcine submandibular gland lacks secondary structure and is thought to assume an extended (nonglobular) solution conformation [15]. Similar findings have been proposed for a 50 amino acid residue tandem repeat of ovine mucin [13] and human salivary mucin MG2 [36]. However, it is important to note that current predictions are only as accurate as the data base used for the comparative analysis [15, 37]; unfortunately, current data bases are composed of 'ideal' globular proteins and thus new rules may need to be formed for more accurate predictions.

*Expression of Apomucin*

At present there is no information regarding the mechanisms which control the expression of salivary apomucins. The subcellular localization of porcine apomucin has been achieved by immunoelectron microscopy [38]; as expected, the apomucin was detected in the rough endoplasmic reticulum (rER), the transitional elements of the rER and the vesicles at the *cis* side of the Golgi apparatus. Colloidal gold coated with *Helix pomatia* lectin was used to localize nonreducing GalNac residues. While no GalNac was observed in the rER, it was heavily concentrated in the *cis* compartment of the Golgi apparatus. Collectively, these data imply that porcine apomucin is synthesized in a manner similar to other secretory proteins and that O-glycosylation is a posttranslational event. Interestingly, the apomucin was detected in specific regions of the secretory apparatus suggesting that there may be microdomains within the secretory circuit. Similar observations have been noted for albumin in hepatocytes [39] and mucin in rat submandibular glands [17]. Alternatively, this biased localization may be the result of fixation artifacts [17, 38]. Clearly, experiments with putative sorting signals will be required to distinguish between the two possibilities.

Recently, cDNA clones encoding a human polymorphic epithelial mucin expressed by carcinomas have been characterized [40, 41]. In common with porcine submandibular gland mucin, the protein core of this species is composed of tandem (20 amino acids) repeats. Southern blot analysis reveals a pattern of two mucin alleles/individual. As suggested above, the number of repeats appears not to be critical for function per se. Thirty different alleles have been determined in 69 random subjects studied thus far.

### *Co- and Posttranslational Modifications of the Apomucin*

Central to understanding the regulation of mucin biosynthesis is delineation of the various co- and posttranslational processes which occur, including glycosylation, acylation, and sulfation. Indirect control is exerted by those factors which either modulate the activity or induce the expression of those enzymes required to effect the co- and posttranslational reactions. The various modifications to the mucin core protein have a profound influence on the overall structure and function of the mucin. Regions of peptide backbone which are not glycosylated are said to be naked and hence more susceptible to proteolytic attack [2]. Compositional studies have indicated that the naked regions of mucins are enriched in both hydrophobic amino acids and cysteine residues [42, 43]. These findings have been supported by fluorescent spectroscopy of hydrophobic binding sites in rat [16, 44] and human [36, 45] salivary mucins. The distribution and extent of naked regions appear to play an important role in dictating the suprastructure of the mucin glycoprotein. Two distinct patterns have emerged. The simpler pattern is characterized by a single mucin peptide core over which oligosaccharides are uniformly distributed. Mucins of this type tend to be smaller in size ($M_r \approx$ 60–200 kd), although they may form large molecular weight suprastructures through noncovalent interactions. Because of the uniform arrangement of oligosaccharides, mucins of this class are resistant to proteolysis (i.e. they have no susceptible naked regions). Examples include human salivary mucin MG2 [6, 7, 36], rat submandibular gland mucin [16, 17], mouse submandibular gland mucin [10, 11], and ovine submandibular gland mucin [12]. In contrast to these simple mucins, higher molecular weight mucins are characterized by a biased distribution of naked regions and complex, covalently stabilized suprastructures. Examples include human salivary mucin MG1 [7, 36], monkey salivary mucin [8], and rat sublingual gland mucin [46]. An understanding of the functional consequences of these divergent structural motifs is beginning to emerge. Classic studies of submandibular gland mucins have repeatedly pointed to the presence of two mucins ('major' and 'minor' [47]) based on affinity for hydroxyapatite (the major inorganic constituent of tooth enamel). Extensive studies with mucins purified from human submandibular-sublingual saliva have indicated that MG2, characterized by uniformly distributed oligosaccharides [6, 7], interacts specifically with the oral bacteria *Streptococcus mitis* and *Streptococcus sanguis* [7, 48]. In contrast, MG1, a higher molecular weight species characterized by a biased distribution of oligosaccharides [7, 36], displays physicochemical properties that enable it to effectively coat the hard and soft tissues of the oral cavity [2]. In vitro studies have supported this view; MG1 has a greater affinity for hydroxyapatite than MG2 [49]. Collectively, the data suggest that MG1 primarily serves a tissue-coating function, thereby protecting the hard and soft tissues, whereas MG2 fails to adsorb in appreciable quantity to the tooth surface, making it available to interact with specific bacteria and effect their clearance from the oral cavity [50]. In this regard, it is interesting to note a recent report in which the relative proportion of MG2 and MG1 in the mucous coat was examined in caries-susceptible and caries-resistant subjects. MG2 predominated in the mucous coat of caries-resistant subjects whereas MG1 and MG2 were present in similar proportion in caries-susceptible subjects [51].

*N-Glycosylation*

The presence of a small number of N-linked oligosaccharides has been reported for mouse submandibular gland mucin [11, 52] and human salivary mucin MG2 [7]. Presumably, the synthesis of these side chains is initiated as a cotranslational event with the 'en bloc' transfer of a $(GlcNac)_2Man_9Glc_3$ oligosaccharide from a dolichol lipid carrier to an asparagine residue on the nascent polypeptide chain [27, 53]. The oligosaccharide undergoes subsequent processing which involves removal of all Glc residues and some Man residues. To form complex linkages, additional sugars such as GlcNac, Fuc and SA are added.

At present, it is not yet known if all mucins contain asparagine-linked oligosaccharides. Moreover, the function of these N-linked carbohydrate side chains is not known. Speculation has centered around a role in intracellular signaling; direct analysis of this possibility should soon be possible through the use of site-directed mutagenesis methods.

*O-Glycosylation*

The bulk of available biochemical and immunocytochemical evidence suggests that O-glycosylation is a posttranslational event that initiates in the *cis* compartment of the Golgi apparatus [17, 38, 53–55] and requires no lipid-bound intermediates [56, 57]. Rather, precursor nucleotide sugars are added stepwise to the growing carbohydrate side chain under the control of specific glycosyltransferases. Thus, an initial control point for the posttranslational modification of a mucin-glycoprotein is the availability of the appropriate sugar nucleotide pool. Sugar nucleotides are produced within the cell cytosol, the exception being the CMP-sialic acid which forms in the nucleus [58]. The translocation of UDP-galactose across the Golgi membrane of the rat submandibular gland, a key sugar precursor in the formation of the RSMG mucin, occurs via facilitated transport [59]. It has been proposed that such transport occurs via antiport proteins which span the Golgi membrane. It is envisaged that the activated sugar binds via the nucleotide to the antiport protein on the cytosolic side and is then subsequently translocated across the membrane, in an equimolar exchange for nucleoside monophosphate in the Golgi lumen [58].

For most mucins, the initial glycosyltransferase which acts on the apomucin is UDP-GalNac: polypeptide α-N-acetylgalactosaminyltransferase. This enzyme catalyzes the addition of a GalNac residue in α-anomeric linkage to either Ser or Thr of the apomucin. As such, the events which regulate the expression and activity of this enzyme are central to controlling mucin biosynthesis. While some progress has been made in the partial purification and characterization of this enzyme from salivary glands [13, 60–62], little is known about its activity in vitro. From comparison of proposed consensus sequences [13, 30, 60–63] it is clear that there is no primary signal embodied within a primary amino acid sequence which directs O-glycosylation. Thus, the presence of Ser or Thr residues are necessary, but not sufficient for O-glycosylation to take place. Perhaps the only common feature to emerge from a comparison of peptide sequences which are O-glycosylated is the prevalence of proline; it has been suggested that this endows the specific segment with a kinkiness which is necessary for substrate accessibility. This has led to the suggestion that peptide secondary structure represents the critical signal, that is, the presence of a Ser or Thr in a β turn would render it more accessible to the glycosyltransferase [13, 30, 64]. However, more recent studies have identified proteins which contain Ser and Thr residues in β turns which fail to become O-glycosylated. For example, we have colocalized the heavily glycosylated rat submandibular gland mucin and nonglycosylated rat glutamine/glutamic acid-rich proteins (GRP) to the same secretory granules of the rat submandibular gland acinar cells [17]. Although the GRP have numerous potential O-glycosylation sites (at least half of which are in predicted β turns) and both GRP and mucin appear to traverse the same intracellular route, no glycosylation occurs in GRP [17]. At present we can only speculate as to why the GRP are not O-glycosylated. One possibility is that the GRP lack an additional molecular signal which is required for O-glycosylation. Alternatively, potential sites may be inaccessible to glycosyltransferases owing to the overall conformation of the GRP or because it forms complexes with other secretory components in the Golgi apparatus.

Other core O-linkages have been identified in eukaryotic cells, for example GlcNac-Ser, Thr [65–67]. The significance of this is not yet understood, but these glycoproteins appear to target to either the endoplasmic reticulum or the nucleus, and in this latter site may function as (or in concert with) *trans*-acting factors in the expression of eukaryotic genes [67].

The elongation of the carbohydrate side chains appears to be dictated by substrate specificity (i.e. the product of the first glycosyltransferase reaction becomes the substrate for the second) [68, 69] and the topographical distribution of the glycosyltransferases themselves in the Golgi apparatus [70]. Implicit within this model is the prediction that there is a unique glycosyltransferase for each sugar added to the mucin. Studies with highly purified glycosyltransferases have supported the one linkage-one enzyme concept [71–73]. Superimposed over the substrate specificity of the transferases is the compartmentalization of the enzymes within the Golgi apparatus itself. Some cell types appear to have a biased distribution of glycosyltransferases within the *cis*, medial and *trans*-Golgi network whereas others do not [70]. Quite clearly, the signals which target specific glycosyltransferases to unique regions of the Golgi need to be identified. Work towards this goal has only just begun, with the successful cloning of several glycosyltransferases including β-galactoside α-2,6-sialyltransferase [74] and β-1,4-galactosyltransferases [75–77]. The organization of these two (conceptually translated) glycosyltransferases is remarkably similar in that each contains three distinct structural domains: (1) a short $NH_2$ terminal domain; (2) a single transmembrane domain (of approximately 20 residues), and (3) a large C terminal domain which has potential sites of glycosylation. The most critical gap in our knowledge related to the glycosyltransferases involves the factors which regulate glycosyltransferase expression and/or activity. Altered glycosyltransferase expression (or activity) in response to hormonal stimulation, development, aging or cellular transformation are but a few of the potential parameters which require future exploration.

*Sulfation*

Structural studies have suggested the presence of GlcNac-6-sulfate in rat salivary [78, 79] and rabbit submandibular gland mucins [80]. GlcNac-4-sulfate has been demonstrated in canine [81] and monkey [82] salivary mucins. Recently, a sulfotransferase enzyme that catalyzes the transfer of a sulfate ester group from 3′-phosphoadenosine-5′-phosphosulfate to the oligosaccharides of a rat sublingual [83] and a rat submandibular [84] gland mucin has been described. Structural characterization of the submandibular gland derived oligosaccharides revealed that the sulfate groups were attached to C-6 of terminal GlcNac residues [85]. The sulfonyltransferase was shown to be highly enriched in the Golgi fraction of the rat submandibular [84] and rat sublingual gland [83]; this localization is consistent with the transfer of sulfate to a preformed oligosaccharide chain. Sulfation in the rat submandibular gland was found to be enhanced in the presence of 16,16-dimethyl-prostaglandin $E_2$ and inhibited by either acetylsalicylic acid [86] or ethanol [87].

The presence of highly charged, sulfate moieties could have a profound impact on the function of salivary mucins [2]. In particular, these anionic groups have been implicated in the interaction of salivary mucins with specific oral bacteria [2, 79], as well as mucosal hydration [2, 87].

*Acylation*

The covalent attachment of fatty acids to mucin-glycoproteins has been reported to occur as both a co- and posttranslational event [88, 89]. It is not yet clear if the same enzyme is involved in co- and posttranslational acylation. Subcellular fractionation of the enzymic activity which catalyzes the transfer of palmitic acid from palmitoyl coenzyme A to the acceptor revealed association with the Golgi-rich membrane fraction [86]. Moreover, there is a discrepancy in the literature concerning the degree to which secreted mucins are acylated in vivo [e.g. cf. 36, 45].

The potential function of fatty acids associated with, or covalently bound to mucin is still open to speculation. For example, it has been suggested that cotranslational regulation could serve as a sorting signal for intracellular routing or as a signal for the O-glycosylation of the mucin. Posttranslational acylation may play a role in the assembly of mucin into high molecular weight polymers [90]. Alternatively, there are considerable data which indicate that acylation of mucin has an impact on the functional role of mucus. For example, mucin-bound fatty acids could contribute significantly to the tissue-coating properties of the salivary mucins [2]. Acylation has been observed to stabilize mucin structure by rendering the molecule more resistant to proteolytic breakdown [44, 91]. Further, the presence of fatty acids results in an enhancement of the ability of mucin to act as a permselective barrier [92, 93]. Differences in the associated and covalently bound lipids have been observed in subjects who are caries-resistant and caries-susceptible [45, 91]. Significantly less lipid is associated with the mucin [45], acquired enamel pellicle [93] and oral mucous coat [51] of caries-resistant subjects than in age-matched caries-susceptible subjects.

The evidence for a genetic predisposition for caries-resistance makes it tempting to speculate that acylation may be at least one genetic factor involved in the host defense against caries.

### *Sorting and Packaging*

Mucins complex with a wide range of peptides and proteins either prior to, or concomitant with, packaging in secretory granules. Studies with mucins from rat [16] and ovine submandibular gland [12], as well as human submandibular-sublingual saliva [6], have shown that small peptides are released after modification of the ε-amino group of lysine. These peptides may be analogous to link proteins that have been described for goblet cell mucins [94, 95]. In addition, a wide range of proteins with protective qualities enter into heterotypic associations with mucins, including antileukoprotease [96], cystatin [97, 98] lysozyme [97, 99], and secretory IgA [6, 97]. These findings lead to the speculation that mucins serve to concentrate protective molecules at the interface of the mucosa and the inhospitable external environment [2, 100]. Thus, these interactions may represent 'prepackaged' associations, which are formulated to express a specific biological function. The intracellular signals which could lead to such associations are not yet understood. However, recent studies point to the importance of the *trans*-Golgi network as a critical depot where sorting of proteins and glycoproteins to secretory granules, lysosomes or the plasma membrane occurs [101].

### *Structural Modulation: Physiological Alteration of Mucin Structure*

Sufficient evidence has accumulated to suggest that the biosynthesis of salivary mucins is subject to a wide range of extrinsic controls. These controls form the basis of a mechanism whereby a cell could modify the structure of a mucin to meet a dynamically changing biological need. We refer to these control mechanisms collectively as structural modulation [2]. Alteration of mucin composition, in response to hormones and neurotransmitters, has been suggested by analysis of either submandibular gland extracts or secretions. For example, pilocarpine stimulation of canine submandibular glands leads to a decrease in the ratio of sialic acid fucose in the secretion [81, 102]. Because these carbohydrates occupy terminal positions on oligosaccharide units, it was suggested that they might compete to terminate a growing carbohydrate chain [102] and/or maintain charge by substitution of sulfate residues for sialic acid moieties [81]. Adrenalectomy or castration of male rats results in a decreased sialic acid content in the submandibular glands [103, 104], whereas spaying female rats had no measurable effect [105]. Chronic isoproterenol administration leads to an increased sialic acid content in the glands [106] of rats and an increased hexose content in the submandibular gland saliva [107]. However, such treatment is now known to induce the expression of an N-linked glycoprotein in the rat submandibular gland [108, 109], perhaps mediated in part via a cAMP-mediated increase in mannosylphosphodolichol synthase activity [110]. These data emphasize the need to characterize structurally the specific molecules of interest from the secretion or glandular extract under study.

The concentrations of a specific mucin have been compared from mouse submandibular gland (ductal) salivary samples elicited by either pilocarpine, epinephrine or isoproterenol [111]. The level of mucin in this secretion was found to be highly stimulus-dependent. At the dosages employed, isoproterenol-stimulated samples contained the highest levels of mucin, whereas pilocarpine-elicited secretions contained the lowest levels. Superimposed over this autonomic control is the diurnal variation observed for mucin content in mice [112]. In keeping with the nocturnal feeding pattern of rodents, mucin concentrations decreased approximately 30% between evening and morning. Confounding the analysis of such data however, was the rather broad variation observed in the mucin concentration among animals.

The influence of either isoproterenol or pilocarpine on the incorporation of carbohydrate in mouse submandibular gland mucins results in an increase in the total amount of sugar incorporated relative to the amino acids; however, the molar ratio of sugars remains unchanged [113]. Relatively few studies have addressed the effect of the normal aging process on mucin biosynthesis per se. A significant difference in the in vitro processing of a rat submandibular gland mucin was observed from tissue derived from aging rats, in comparison with similar material from young rats [114]. Clearly, additional studies are warranted in this area.

## *Future Directions*

Currently we lack the most fundamental information concerning the organization and number of genes which encode the apomucin. An obvious goal (and one which many laboratories are actively pursuing) will be to isolate and characterize genomic clones which encode mucin protein core. From these studies will be derived basic information regarding the *cis*-acting region of DNA which are important in regulating the transcription of the mucin protein core. Ultimately, *trans*-acting factors may be identified which serve to modulate these transcriptional activities.

Another fundamental area which has not yet been explored is the extent to which mucin exocytosis and gene transcription of the apomucin are coupled. Many of the inducible factors which modulate gene transcription are known to act as intracellular messengers in the exocytotic process. The intrinsic controls which regulate the overall interrelational modulatory processes will require elucidation before a complete understanding of mucin biosynthesis is achieved.

## *References*


1 Nugent J, O'Connor M (eds): Mucus and mucosa. CIBA Symp 1984;109:1–256.
2 Tabak LA, Levine MJ, Mandel ID, et al: Role of salivary mucins in the protection of the oral cavity. J Oral Pathol 1982;11:1–17.
3 Mandel ID: The functions of saliva. J Dent Res 1987;66:623–627.
4 Fox PC, Van Der Ven PF, Sonies BC, et al: Xerostomia: evaluation of a symptom with increasing significance. J Am Dent Assoc 1985;110:519–525.
5 Levine MJ, Aguirre A, Hatton M, et al: Artificial salivas: present and future. J Dent Res 1987;66:693–698.
6 Prakobphol A, Levine MJ, Tabak LA, Reddy MS: Purification of a low molecular weight mucin-type glycoprotein from human submandibular-sublingual saliva. Carbohydr Res 1982;108:111–122.
7 Levine MJ, Reddy MS, Tabak LA, et al: Structural aspects of salivary glycoproteins. J Dent Res 1987;66:436–441.
8 Herzberg MC, Levine MJ, Ellison SA, et al: Purification and characteristics of a monkey salivary mucin. J Biol Chem 1979;254:1487–1494.
9 Tabak LA, Reddy MS, Levine MJ: Characterization of a pentasaccharide in salivary mucin from the stumptail monkey *Macaca arctoides*. Arch Oral Biol 1982;27:297–303.
10 Denny PA, Denny PC: Purification and biochemical characterization of a mouse submandibular sialomucin. Carbohydr Res 1980;87:265–274.
11 Denny PA, Denny PC: A mouse submandibular sialomucin containing both N- and O-glycosidic linkages. Carbohydr Res 1982;110:305–314.
12 Hill HD Jr, Reynolds JA, Hill RL: Purification, composition, molecular weight, and subunit structure of ovine submaxillary mucin. J Biol Chem 1977;252:3791–3798.
13 Hill HD Jr, Schwyzer M, Steinman HM, et al: Ovine submaxillary mucin. Primary structure and peptide substrates of UDP-N-acetylgalactosamine: Mucin transferase. J Biol Chem 1977;252:3799–3804.

14 Carlson DM: Structures and immunochemical properties of oligosaccharides isolated from pig submaxillary mucins. J Biol Chem 1987;243:616–626.
15 Eckhardt AE, Timpte CS, Abernethy JL, et al: Structural properties of porcine submaxillary gland apomucin. J Biol Chem 1987;262:11339–11344.
16 Tabak LA, Mirels M, Monte LD, et al: Isolation and characterization of a mucin-glycoprotein from rat submandibular glands. Arch Biochem Biophys 1985;242:383–392.
17 Moreira JE, Tabak LA, Bedi GS, et al: Light and electron microscopic immunolocalization and rat submandibular gland mucin-glycoprotein and glutamine/glutamic acid-rich proteins. J Histochem Cytochem 1989;37:515–528.
18 Gerken TA, Dearborn DG: Carbon-13 NMR studies of native and modified ovine submaxillary mucin. Biochemistry 1984;23:1485–1497.
19 Rose MC, Voter WA, Sage H, et al: Effects of deglycosylation on the architecture of ovine submaxillary mucin glycoprotein. J Biol Chem 1984;259:3167–3172.
20 Shogren RL, Jamieson AM, Blockwell J, et al: Conformation of mucous glycoproteins in aqueous solvents. Biopolymers 1986;25:1505–1517.
21 Gerkin TA: The solution structure of mucous glycoproteins: Proton NMR studies of native and modified ovine submaxillary mucin. Arch Biochem Biophys 1986;247: 239–253.
22 Shogren RL, Jentoft N, Gerken TA, et al: Light-scattering studies of fractionated ovine submaxillary mucins. Carbohydr Res 1987;160:317–327.
23 Gerken TA, Jentoft N: Structure and dynamics of porcine submaxillary mucin as determined by natural abundance carbon-13 NMR spectroscopy. Biochem 1987;26: 4689–4699.
24 Sellers LA, Allen A, Morris ER, et al: Submaxillary mucins. Intermolecular interactions and gel-forming potential of concentrated solutions. Biochem J 1988;256:599–607.
25 Quissel DO, Tabak LA: Salivary mucin secretion; in Handbook of Physiology. Bethesda, Md., American Physiological Society, 1989, vol 111, pp 79–91.
26 Schachter H: Biosynthetic controls that determine the branching and microheterogeneity of protein-bound oligosaccharides. Biochem Cell Biol 1986;64:163–181.
27 Gottschalk A: Biosynthesis of glycoproteins and its relationship to heterogeneity. Nature (Lond) 1969;222:452–454.
28 Gallagher JT, Cornfield AP: Mucin-type glycoproteins: New perspectives on their structure and synthesis. Trends Biochem Sci 1978;3:38–41.
29 Pigman W, Moschera J, Weiss M, et al: The occurrence of repetitive glycopeptide sequences in bovine submaxillary glycoprotein. Eur J Biochem 1973;32:148–154.
30 Timpte CS, Eckhardt AE, Abernethy JL, et al: Porcine submaxillary gland apomucin contains tandemly repeated identical sequences of 81 residues. J Biol Chem 1988; 263:1081–1088.
31 Ann DK, Clements S, Johnstone EM, et al: Induction of tissue-specific proline-rich protein multigene families in rat and mouse parotid glands by isoproterenol. J Biol Chem 1987;262:899–904.
32 Garfinkle MD, Pruitt RE, Meyerowitz EM: DNA sequences, gene regulation and modular protein evolution in the Drosophila 68C glue gene cluster. J Mol Biol 1983; 168:765–789.
33 Dickinson DP, Mirels L, Tabak LA, et al: Rapid evolution of variants in a rodent multi-gene family encoding salivary proteins. Mol Biol Evol 1989;6:80–102.

34 Heinrich G, Habener JF: Genes encoding proteins with homologous contiguous repeat sequences are highly expressed in the serous cells of the rat submandibular gland. J Biol Chem 1987;262:5260–5270.
35 Mirels L, Bedi GS, Dickinson DP, et al: Molecular characterization of glutamine/glutamic acid-rich secretory proteins from rat submandibular glands. J Biol Chem 1987;262:7289–7297.
36 Loomis RE, Prakobphol A, Levine MJ, et al: Biochemical and biophysical comparison of the two mucins from human submandibular and sublingual saliva. Arch Biochem Biophys 1987;258:452–464.
37 Tabak LA, Bowen WH: The roles of saliva (pellicle), diet, and nutrition on plaque formation. J Dent Res 1989;68:1560–1566.
38 Deschuyteneer M, Eckhardt AE, Roth J, et al: The subcellular localization of apomucin and non-reducing terminal N-acetylgalactosamine in porcine submaxillary glands. J Biol Chem 1988;263:2452–2459.
39 Yokota S, Fahimi HD: Immunocytochemical localization of albumin in the secretory apparatus of rat liver parenchymal cells. Proc Natl Acad Sci USA 1981;78:4970–4974.
40 Gendler SJ, Burchell JM, Duhig T, et al: Cloning of partial cDNA encoding differentiation and tumor-associated mucin glycoproteins expressed by human mammary epithelium. Proc Natl Acad Sci USA 1987;84:6060–6064.
41 Gendler SJ, Taylor-Papadimitriou J, Duhig T, et al: A highly immunogenic region of human polymorphic epithelial mucin expressed by carcinomas is made up of tandem repeats. J Biol Chem 1988;263:12820–12823.
42 Donald ASR: The products of pronase digestion of purified blood group-specific glycoproteins. Biochim Biophys Acta 1973;317:420–436.
43 Shogren RL, Jamieson AM, Blackwell J, et al: The thermal depolymerization of porcine submaxillary mucin. J Biol Chem 1984;259:14657–14662.
44 Slomiany BL, Murty VLN, Sarocick J, et al: Role of associated and covalently bound lipids in salivary mucin hydrophobicity: Effect of proteolysis and disulfide bridge reduction. Biochem Biophys Res Commun 1988;151:1046–1053.
45 Slomiany BL, Murty VLN, Slomiany A, et al: Mucus glycoproteins of human saliva: Differences in the associated and covalently bound lipids with caries. Biochim Biophys Acta 1986;882:18–28.
46 Moschera J, Pigman W: The isolation and characterization of rat sublingual mucus-glycoprotein. Carbohydr Res 1975;40:53–67.
47 Gottschalk A, Bhargava AS, Murty VLN: Ovine submaxillary mucins; in Gottschalk A (ed): Glycoproteins. New York, Elsevier, 1972, vol 5, pp 810–817.
48 Murray PA, Levine MJ, Tabak LA, et al: Specificity of salivary-bacterial interactions. II. Evidence for a lectin on *Streptococcus sanguis* with specificity for a NeuAc α2,3Galβ1,3GalNac sequence. Biochem Biophys Res Commun 1984;106:390–396.
49 Tabak LA, Levine MJ, Jain NK, et al: Adsorption of human salivary mucins to hydroxyapatite. Arch Oral Biol 1985;30:423–427.
50 Levine MJ, Tabak LA, Reddy MS, et al: Nature of salivary pellicles in microbial adherence: Role of salivary mucins; in Mergenhagen SE, Rosan B (eds): Molecular Basis of Oral Microbial Adhesion. Washington, American Society of Microbiology, 1985, pp 125–130.
51 Slomiany BL, Murty VLN, Mandel ID, et al: Mucus glycoproteins and lipids of human oral mucosal mucus coat: Differences in physicochemical characteristics with caries. Arch Oral Biol 1989;34:229–237.

52 Nieuw Amerongen AV, Oderkerk, CH, Roukema PA, et al: Murine submandibular mucin (MSM); a mucin carrying N- and O-glycosylically bound carbohydrate-chains. Carbohydr Res 1983;115:C1–C5.
53 Hanover JA, Lennarz WJ: Transmembrane assembly of membrane and secretory glycoproteins. Arch Biochem Biophys 1981;211:1–19.
54 Hanover JA, Elting J, Mintz GR, et al: Temporal aspects of the N- and O-glycosylation of human chorionic gonadotropin. J Biol Chem 1982;257:10172–10177.
55 Johnson DC, Spear PG: O-linked oligosaccharides are acquired by herpes simplex virus glycoproteins in the Golgi apparatus. Cell 1983;32:987–997.
56 Babczinski P: Evidence against the participation of lipid intermediates in the in vitro biosynthesis of serine-(threonine)-N-acetyl-*D*-galactosamine linkages in submaxillary mucin. FEBS Lett 1980;117:207–211.
57 Hanover JA, Lennarz WJ, Young J: Synthesis of N- and O-linked glycopeptides in oviduct membrane preparations. J Biol Chem 1980;255:6713–6716.
58 Hirschberg CB, Snider MD: Topography of glycosylation in the rough endoplasmic reticulum and Golgi apparatus. Annu Rev Biochem 1987;56:63–87.
59 Barthelson R, Roth S: Topology of UDP-galactose cleavage in relation of N-acetyllactosamine formation in Golgi vesicles. Biochem J 1985;225:67–75.
60 McGuire EJ, Roseman S: Enzymatic synthesis of the protein-hexosamine linkage in sheep submaxillary mucin. J Biol Chem 1967;242:3745–3755.
61 Young JD, Tsuchiya D, Sandlin DE, et al: Enzymic O-glycosylation of synthetic peptides from sequences in basic myelin protein. Biochemistry 1979;18:4444–4448.
62 Hogopian A, Westall FC, Whitehead JS, et al: Glycosylation of the A1 protein from myelin by a polypeptide N-acetylgalactosaminyltransferase. J Biol Chem 1971;246:2519–2523.
63 Shimanura M, Inove Y, Inove S: Evidence for unique homologous peptide sequences around the glycosylated seryl and threonyl residues in polysial glycoproteins isolated from the unfertilized eggs of the Pacific salmon *Oncorhynchus keta*. Biochemistry 1985;24:5470–5480.
64 Aubert JP, Biserte G, Loucheux-Lefebure MH: Carbohydrate-peptide linkage in glycoproteins. Arch Biochem Biophys 1976;175:410–418.
65 Holt GD, Snow CM, Senier A, et al: Nuclear pore complex glycoproteins contain cytoplasmically disposed O-linked N-acetylglucosamine. J Cell Biol 1987;104:1157–1164.
66 Abeijon C, Hirschberg CB: Intrinsic membrane glycoproteins with cytosol-oriented sugars in the endoplasmic reticulum. Proc Natl Acad Sci USA 1988;85:1010–1014.
67 Jackson SP, Tjian R: O-glycosylation of eukaryotic transcription factors: Implications for mechanisms of transcriptional regulation. Cell 1988;55:125–133.
68 Roseman S: The synthesis of complex carbohydrates by multi-glycosyltransferase systems and their potential function in intercellular adhesion. Chem Phys Lipids 1970;5:270–297.
69 Watkins WM: Glycosyltransferases. Early history, development and future prospects. Carbohydr Res 1986;149:1–12.
70 Roth J: Subcellular organization of glycosylation in mammalian cells. Biochim Biophys Acta 1987;906:405–436.
71 Beyer TA, Rearick JI, Paulson JC, et al: Biosynthesis of mammalian glycoproteins. J Biol Chem 1979;254:12531–12541.

72 Williams D, Schachter H: Mucin synthesis. I. Detection in canine submaxillary glands of an N-acetylglucosaminyltransferase which acts on mucin substrate. J Biol Chem 1980;255:11247–11252.
73 Williams D, Longmore G, Matta KL, et al: Mucin synthesis. II. Substrate specificity and product identification studies on canine submaxillary gland UDP-GlcNac:Gal-β-1-3GalNac(GclNac→GalNac)-β-6-N-acetylglucosaminyltransferase. J Biol Chem 1980;255:11253–11261.
74 Weinstein J, Lee EV, McEntee K, et al: Primary structure of β-galactoside α-2,6-sialyltransferase. J Biol Chem 1987;262:17735–17743.
75 Shaper NC, Hollis GF, Douglas JG, et al: Characterization of the full-length cDNA for murine β-1,4-galactosyltransferase. J Biol Chem 1988;263:10420–10428.
76 Nakazawa K, Ando T, Kimura T, et al: Cloning and sequencing of a full-length cDNA of mouse N-acetylglucosaminide (β-1,4)galactosyltransferase. J Biochem 1988;104: 165–168.
77 Masri KA, Appert HE, Fukuda MN: Identification of the full-length coding sequence for human galactosyltransferase (β-N-acetylglucosaminide: β-1,4-galactosyltransferase). Biochem Biophys Res Commun 1988;157:657–663.
78 Embery G, Ward C: The nature of the sulphate grouping in a rat salivary sulphated glycoprotein. Arch Oral Biol 1976;21:627–629.
79 Green DRJ, Embery G: Isolation, chemical and biological characterization of sulphated glycoproteins synthesized by rat buccal and palatine minor salivary glands in vivo and in vitro. Arch Oral Biol 1987;32:391–399.
80 Hofsoy H, Jonsen J: In vivo $^{35}SO_4$-labeling of rabbit salivary components. Arch Oral Biol 1987;32:391–399.
81 Lombart CG, Winzler RJ: Isolation and characterization of oligosaccharides from canine submaxillary mucin. Eur J Biochem 1974;45:77–86.
82 Tabak LA, Levine MJ: Characterization of sulphated monosaccharides from stump-tail monkey salivary mucin. Arch Oral Biol 1981;26:315–317.
83 Slomiany A, Murty VLN, Liau YH, et al: Enzymatic sulphation of mucin glycoprotein in rat sublingual salivary gland. Arch Oral Biol 1988;33:669–676.
84 Slomiany B, Murty VLN, Sarosiek J, et al: Enzymatic sulfation of mucus glycoprotein in rat submandibular salivary gland. Int J Biochem 1985;21:165–171.
85 Slomiany B, Piotrowski J, Nishikawa H, et al: Enzymatic sulfation of salivary mucins: Structural features of $^{35}S$-labelled oligosaccharides. Biochem Biophys Res Commun 1988;157:61–67.
86 Murty VLN, Bilski J, Liau YH, et al: Sulfation in vitro of mucus glycoprotein by submandibular salivary gland: Effects of prostaglandin and acetylsalicylic acid. Biochim Biophys Acta 1988;966:287–296.
87 Slomiany BL, Liau YH, Zalesna G, et al: Effect of ethanol on the in vitro sulfation of salivary mucin. Alcohol Clin Exp Res 1988;12:1–6.
88 Slomiany A, Mizuta K, Zalesna G, et al: Co-translational processing and intracellular transport of rat salivary mucus glycoprotein. Arch Oral Biol 1988;33:807–818.
89 Slomiany BL, Liau YH, Carter SR, et al: Enzymatic acylation of mucus glycoprotein with palmitic acid in rat submandibular salivary gland. Arch Oral Biol 1986;31:463–468.
90 Slomiany BL, Murty VLN, Takagi A, et al: Fatty acid acylation of salivary mucin in rat submandibular glands. Arch Biochem Biophys 1985;242:402–410.
91 Slomiany BL, Murty VLN, Carter SR, et al: Susceptibility of salivary mucin to proteolysis: Differences with caries. Ann NY Acad Sci 1987;494:356–358.

92 Sarosiek J, Slomiany A, Takagi A, et al: Hydrogen ion diffusion in dog gastric mucus glycoprotein: Effect of associated lipids and covalently bound fatty acids. Biochem Biophys Res Commun 1981;118:523–531.
93 Slomiany BL, Murty VLN, Zdebska E, et al: Tooth surface-pellicle lipids and their role in the protection of dental enamel against lactic acid diffusion in man. Arch Oral Biol 1986;31:187–191.
94 Mantle M, Forstner GG, Forstner JF: Antigenic and structural features of goblet-cell mucin of human small intestine. Biochem J 1984;217:159–167.
95 Fahim REF, Forstner GG, Forstner JT: Heterogeneity of rat goblet-cell mucin before and after reduction. Biochem J 1983;209:117–124.
96 Schill WB, Wallner O, Schiessler H, et al: Immunofluorescent localization of the acid-stable proteinase inhibitor (antileukoprotease) of human cervical mucus. Experientia 1978;34:509–510.
97 Shomers JP, Tabak LA, Levine MJ, et al: The isolation of a family of cysteine-containing phosphoproteins from human submandibular-sublingual saliva. J Dent Res 1982;61:973–977.
98 Al-Hashimi I, Dickinson DP, Levine MJ: Purification, molecular cloning and sequencing of salivary cystatin SA-I. J Biol Chem 1988;263:9381–9387.
99 Creeth J, Bridge JL, Horton JR: An interaction between lysozyme and mucus glycoproteins. Biochem J 1979;181:717–724.
100 Edwards PAW: Is mucus a selective barrier to macromolecules. Br Med Bull 1978; 34:55–56.
101 Grossman A: Information transfer in biological systems: Targeting of proteins to specific organelles onto the extracellular environment (secretion). Comp Biochem Physiol 1988;191B:389–424.
102 Dische Z, Burgher CM, Danilchenko A, et al: Variations in the composition of glycoproteins of the submaxillary and parotid saliva of dog in relation to intensity of secretory stimulus. Arch Biochem Biophys 1969;135:1–9.
103 Curbelo HM, Camper CH, Kofoed JA, et al: Effects of testosterone propionate on sialic acid content of submaxillary and retrolingual glands in the rat. J Dent Res 1974;53:1164–1166.
104 Curbelo HM, Houssay AB, Camper CH, et al: Influence of the adrenal glands on the sialic acid content of submaxillary and retrolingual glands in the rat. J Dent Res 1973;52:1265–1276.
105 Curbelo HM, Houssay AB, Camper CH, et al: Effects of oestrogens upon the sialic acid in the submaxillary and sublingual glands in the rat. Arch Oral Biol 1974;19: 421–423.
106 Curbelo HJ, Devalle JJ, Houssay AB, et al: Effects of isoproterenol upon the sialic acid content of salivary glands in the rat. J Oral Ther Pharmacol 1968;4:431–438.
107 Zengo AN, Mandel ID, Solomon A, et al: Chemistry of rat saliva following isoproterenol administration. J Oral Ther Pharmacol 1968;4:359–363.
108 Mehansho H, Carlson DM: Induction of protein and glycoprotein synthesis in rat submandibular glands by isoproterenol. J Biol Chem 1983;258:6616–6620.
109 Humphreys-Beher MG: Isoprenaline-induced changes in rat parotid and submandibular glands are age- and dose-dependent. Biochem J 1984;221:15–20.
110 Banerjee DK, Kousvelari EE, Baum BJ: cAMP-mediated protein phosphorylation of microsomal membranes increases mannosyl phosphodolichol synthase activity. Proc Natl Acad Sci USA 1987;84:6389–6393.

111 Denny PC, Denny PA, Yim MS: The effects of various secretagogues on the mucin content of pure submandibular salivas. J Dent Res 1987;66:1011–1015.
112 Denny PC, Denny PA: Diurnal variation of sialomucin concentration in female mouse submandibular glands measured by radioimmunoassay. Arch Oral Biol 1984; 29:1033–1040.
113 Nieuw Amerongen AV, Marjam EG, Peternel A, et al: Influence of autonomic agonists on the in vivo incorporation of [$^{3}$C]Leveine and N-acetyl[$^{14}$C]mannosamine into submandibular mucin of the mouse. Biochim Biophys Acta 1984;798:103–110.
114 Baum BJ, Kuyatt BL, Humphreys S: Protein production and processing in young adult and aged rat submandibular gland cells in vitro. Mech Aging Dev 1983;23: 123–136.

L.A. Tabak, DDS, PhD, Departments of Dental Research and Biochemistry,
University of Rochester, Rochester, NY 14642 (USA)

Ferguson DB (ed): Aspects of Oral Molecular Biology.
Front Oral Physiol. Basel, Karger, 1991, vol 8, pp 95–116

# Genetic Control and Structure of the Proline-Rich Proteins of Human Saliva

*D.B. Ferguson*

Department of Physiological Sciences, University of Manchester, UK

In 1965, Mandel et al. [1] reported that a number of salivary proteins contained unusually large amounts of proline. Since the first of these was isolated and characterized by Bennick [2] in 1977, there has been intensive study of their structure, inheritance patterns and, most recently, the genetic control of their production. Four major reviews, covering this subject area from different angles, have appeared in the last 3 years [3–6].

The proline-rich proteins make up a large proportion of the protein content of saliva – a figure as high as 70% was deduced by Kauffman and Keller [7], although their method was one subject to quite wide margins of error. In parotid saliva, apart from the isozymes of amylase, lysozyme, and small amounts of albumin, statherin, the histidine-rich proteins, immunoglobulins, peroxidase, lipase and minor enzymes, all the remaining proteins stain as proline-rich proteins. This includes the glycoproteins.

Three broad groups of proteins have been distinguished by their behavior on electrophoresis and their staining properties: the acidic proline-rich proteins, the basic proline-rich proteins, and the glycosylated proline-rich proteins. In all of these, the three amino acids, proline, glycine and glutamine, make up 70–88% of the total amino acid residues and proline alone is around 20–40% [8]. In the one subject studied by Kauffman and Keller [7] the proportion of acidic:basic:glycosylated was 5:4:3.

So far, these proteins appear to be almost unique to saliva, having otherwise been found only in the respiratory tract [9, 10] and pancreas [11]. It is less easy to assign specific functions to them. All of them bind tannins [12], a property which may reduce the ability of tannins to bind to food materials and inhibit their absorption in the small intestine. The acidic proline-rich proteins usually contain phosphoserine; they bind calcium [13], maintaining salivary calcium levels and inhibiting the formation of hydroxyapatite, thus decreasing the possibility of formation of dental calculus [14]. In similar fashion they bind to hydroxyapatite surfaces, constituting part of the salivary pellicle on dental enamel [15] and then mediating adherence of bacteria to the pellicle and enamel surface in the formation of dental plaque [16]. The glycosylated proline-rich proteins act predominantly as lubricants, but may also bind to some microorganisms and have a role in the microbial ecology of the mouth [17]. Although there is no direct evidence linking proline-rich proteins with dental caries experience, a greater increase in caries has been demonstrated in children inheriting some combinations of proline-rich proteins. It seems more likely, however, that caries experience is related to the peroxidase variants in saliva which are correlated with some of the acidic proline-rich proteins [18].

The first acidic proline-rich proteins to be studied were designated proteins A and C by Bennick's group [2, 19], but these were later subsumed into a broader group of PRP1, PRP2, PRP3, and PRP4 as numbers 3 and 1 [20]. Hay and Oppenheim [21] described a number of other minor acidic proline-rich proteins which were structurally similar to the main four but had different electrophoretic mobilities. Bennick [22] suggested that these probably had differing degrees of amidation of amino acids or phosphorylation of the serine. Azen and Denniston [23] described a protein existing as a double band on electrophoresis, which they named Db, and Friedman and Merritt [24] added another named Pa. More recently, PIF (parotid isoelectric focusing variant, another double-banded protein) [25] and As (a variant identified in sodium dodecyl sulphate (SDS) gels) [26] have been described. It should be commented that at least four types of acrylamide gel electrophoresis have been used in these identifications – acidic, alkaline, isoelectric focusing, and SDS – and some cross-referencing by two-dimensional gels might clarify the relationships of the bands staining for proline-rich proteins. This has been attempted by comparing the positions of purified proteins in different gels by Azen [3]. The electrophoretic data available at present are summarized in figures 1 and 2.

The basic proline-rich proteins have been studied by a number of groups. Kauffman et al. [27–29] isolated nine basic proline-rich proteins which they named IB-1 to IB-9. Isemura et al. [30–32] and Saitch et al. [33–35] also described nine proteins but named them BPRP or P-A to I. Relationships between the two classifications are complex and further compounded by

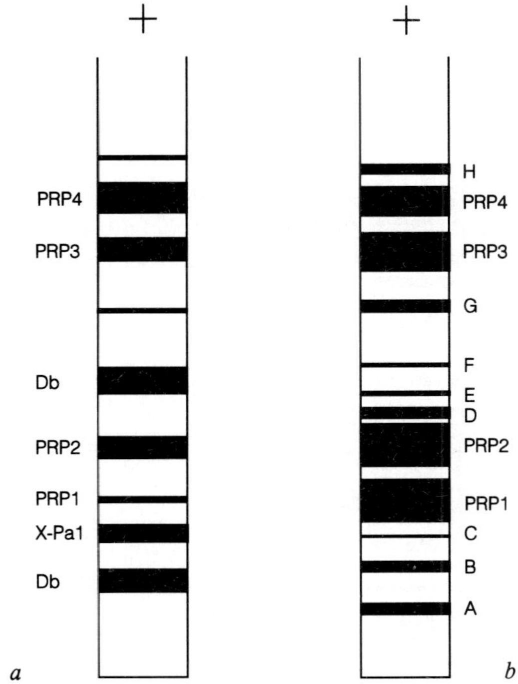


*Fig. 1.* Electrophoresis of salivary proteins in alkaline polyacrylamide gels. *a* A composite drawn from Azen [69]. *b* A composite of the patterns from fractions derived from gel filtration by Hay and Oppenheim [21]. The letters indicate the fractions they described at that time.

description and naming of similar proteins by other workers. Thus, Isemura's P-E is identical with Kauffman and Keller's IB-9. Anderson et al. [36] identified a protein PmS which was identical with IB-6. Another group of proteins with five variants, the Ps system, showed racial variation [6, 37–39]. Other basic proline-rich proteins include Po [40], Pc [41], Pe [40], three proteins Pmo [26, 42] and P3 [26].

The glycosylated proteins which have been identified include eight G1 proteins, named from the G1 locus which controls their synthesis (four by Azen et al. [43] and a further four by Minaguchi et al. [44], and two proteins designated CON1 and CON2 which contain alpha-*D*-mannose and can be detected by their binding of the lectin concavalin [45].

All the proteins which have so far been given designations are listed in table 1.

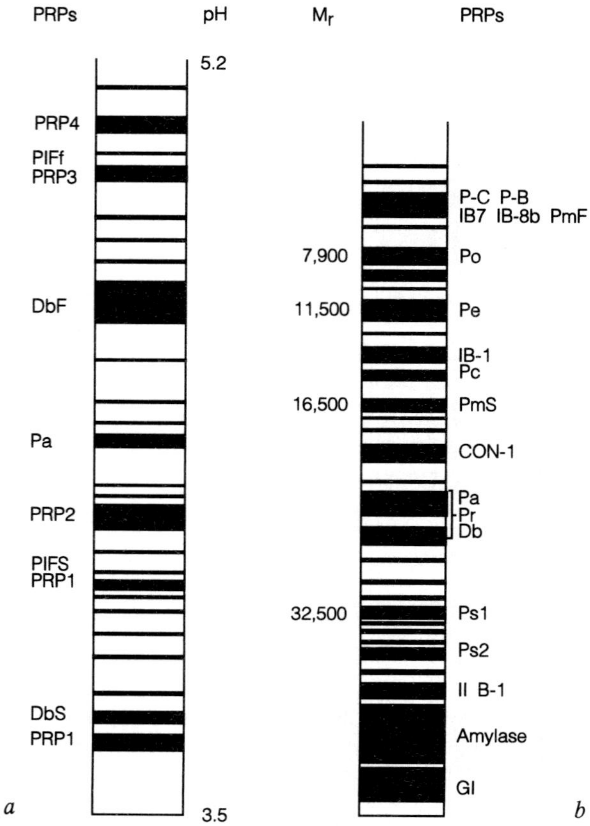


*Fig. 2.* Other electrophoretic separations of salivary proline-rich proteins. *a* Isoelectric focusing gels: a composite diagram from gels shown by Azen [3]. *b* SDS-PAGE: a composite diagram from gels shown by Azen [3]. These diagrams would be equivalent to pooled saliva samples.

## *Structure of the Proline-Rich Proteins*

All the proline-rich proteins contain large amounts of proline, glycine and glutamine and it is perhaps not too surprising that these amino acids are arranged in common repeating groups [46]. In the acidic proline-rich proteins this was characterized as PRO-GLN-GLY-PRO-PRO-GLN-GLY-GLY (PQGPPQQGG in the one-letter amino acid codes) [22] although a longer repeat with small variations could be identified which included this segment.

*Table 1.* Proline-rich proteins identified in parotid saliva

| Acidic PRP | Basic PRP | Glycosylated PRP |
|---|---|---|
| PRP1* protein A Pr1 | IB-1* | GI-1 |
| PRP2* Pr2 | IB-2* | GI-2 |
| PRP3* protein C Pr3 | IB-3* | GI-3 |
| PRP4* Pr4 | IB-4* | GI-4 |
| DB f and s** | IB-5* | GI-5 |
| Pa** | IB-6* PmS | GI-6 |
| PIF f and s** | IB-7* | GI-7? |
| As | IB-8* | GI-8? |
| Minor components A-H | IB-9* P-E PmF | CON1 |
| | Ps | CON2 |
| | Pe | DEAE II-1 |
| | Po | |
| | Pc | |
| | Pmo1 | |
| | Pmo2 | |
| | Pe | |
| | P-A* | |
| | P-B* | |
| | P-C* | |
| | P-D* | |
| | P-F* | |
| | P-G | |
| | P-H* | |
| | P-I | |
| | DEAE II-2* | |

A glycosylated proline-rich basic protein, HSLSM-BPRP, has been characterized in mixed submandibular/sublingual secretion [71]. See text for individual references.
*Indicates that the protein has been sequenced; **indicates that the amino acid sequence has been deduced from nucleotide sequences; ?indicates that these were probably other proteins already described.

This is shown in figure 3. Later work from the genetic point of view identified a repeat (the H or HAEIII repeat) [47]:

PRO-PRO-GLN-GLY-LYS/ARG-PRO-GLN-GLY-PRO-PRO-GLN-GLN-GLY-GLY-HIS-(GLN-GLN-GLY-PRO)-PRO-PRO or PPQGK/RPQGPPQQGGHQ/.Q/.G/.P/.PP

which includes the above sequence and extends it. The / denotes an alternative amino acid residue and the bracketed section or /. indicates that these amino acids may be omitted from the repeat. In the acidic proline-rich

N1

E S L A
Q D L N E D U S Q E P L I
D U U S

1 5 10 15 17

H

K Q G Q P
P P Q G P Q G P P Q Q G G H P P
R – – – –

1 5 10 15 17 21

PRP3

N1

S* D U U S*

1 8 11 12 15 17

N2

D G G D S*E Q F I D E E R Q G P P L G G Q Q S Q P S A G D G N Q N D

18 20 25 30 35 40 45 50

– Q Q Q G

52 53 54 55 56 62 63 64 65 67 69

H 4 5 6 7 8 14 15 16 17 19 21

K – – – –

70 74 84 85 86

H 1 5 15 16 17 18 19 20 21

R – – – –

87 91 101 103

H 1 5 15 16 17 18 19 20 21

R R

104 106 108 124

H 1 3 5 21

P K P R P –

125 127 127 135 139 140 142 143 144

H 1 3 5 11 15 16 17 18 19 20 21

Q G Q S P Q

proteins two N-terminal sequences are present, one being common with the N-terminal sequence of the basic and glycosylated proline-rich proteins, and the other conferring the acidic nature of the proteins. This second N-terminal sequence contains two phosphoserine residues, and it is this section which confers the property of calcium binding on the molecules. The C-terminal sequence of the acidic proteins also appears to differ from those of the basic and glycosylated proteins [47].

The characteristic repeat of the basic proline-rich proteins was identified by Bennick [4] on a basis of the reports of Kauffman et al. [27, 28] and Isemura et al. [30–32] as PRO-GLN-GLY-PRO-PRO-PRO-GLN-GLY-GLY (PQGPPPQGG). Working from cloned DNA, Azen et al. [47] obtained consensus sequences for repeat regions in the basic and glycosylated proline-rich proteins. They suggested the following sequence as the most common:

PRO-PRO-PRO-GLN-LYS-PRO-GLN-GLY-PRO-PRO-PRO-GLN-GLN-ASN-LYS-PRO-GLN-GLY-PRO-PRO

At least three minor variants of this were also suggested, two of them with Gln instead of the second Lys (fig. 4). All three of the variants included serine, and these were found particularly in the glycosylated proteins.

The acidic proline-rich proteins were the first to be isolated and characterized. They contain 106–171 amino acid residues and range in molecular weight from about 10,000 to 16,000 daltons. Wong et al. [48] determined the complete sequence for Bennick's protein C and found it to be the same as the first 106 residues of protein A (fig. 3). This confirmed the original hypothesis of Friedman and Karn [49] and the initial work of Karn et al. [50]. Protein C and protein A in turn were found to be identical with PRP1 and PRP3 described by Oppenheim et al. [20]. The other two proteins isolated by Oppenheim et al. [20], PRP2 and PRP4, differed from them only in having aspartic acid instead of asparagine in position 4 [8, 51]. Some confusion had

---

*Fig. 3.* Sequence of amino acids in N1, the N-terminal sequence deduced by Maeda et al. [70] from cloning studies, in H, the repeat sequence of the acidic proline-rich proteins similarly deduced, and in PRP3 from the determination by Wong et al. [48]. Differences from the deduced sequences are shown by a – for a missing amino acid, the appropriate letter for a different amino acid, and the insertion of an extra amino acid by its letter without a corresponding number in that block. *Indicates a phosphate group on the serine. Each block in the protein is numbered first with the protein sequence numbers, and then below that with the numbers of the deduced block sequence.

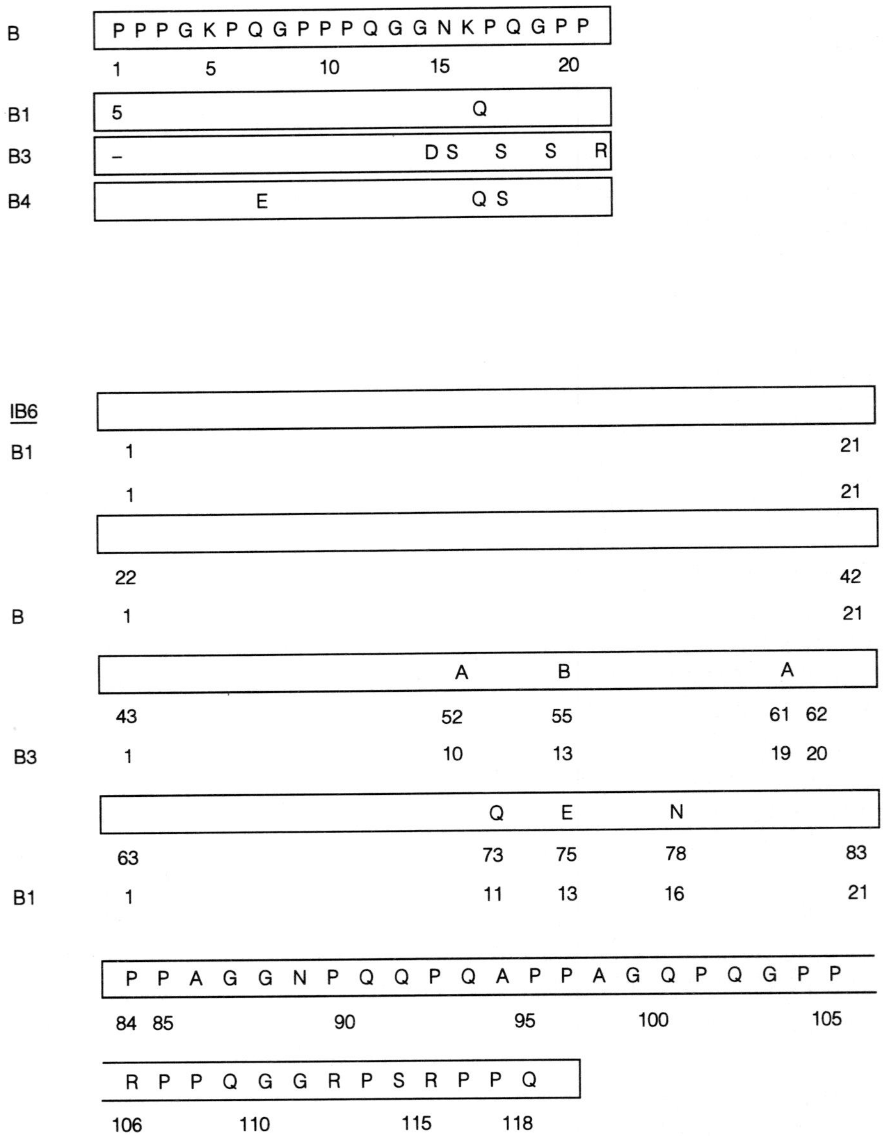


*Fig. 4.* Amino acid sequence of the basic proline-rich proteins: B, B1, B3 and B4 are the repeat sequences deduced by Maeda et al. [70] from their cloning studies, and the structure of IB6, as determined by Kauffman et al. [27] is shown as the repeat units. Differences from the deduced sequence are indicated in the blocks by the different amino acid.

arisen over the structure of protein A and protein C because earlier determinations were carried out with samples contaminated with PIFs and PIFf which have Asn and Asp at positions 4 and 50. Corrected sequences for PRP1, PRP2, PRP3 and PRP4 have now been published by Hay et al. [52] and agree with those put forward by Azen et al. [53]. The relationship between PRP1 and PRP3 became clearer when Wong et al. [54] demonstrated that PRP1 could be split by salivary kallikrein at the arginine-glycine bond at position 106 to yield PRP3, and PRP2 could similarly yield PRP4. A peptide of 44 amino acid residues from the C-terminal ends was termed P-C. The other acidic proline-rich proteins, with the exception of Pa, all occur as double bands on electrophoresis, and it is likely that this reflects the splitting of the proteins in a similar fashion to that described. PIF slow protein [52] is similar to PRP1 with asparagine at position 4 and aspartic acid at position 50 instead of the reverse. Azen et al. [53] cloned and sequenced the exon and intron structures of the genes for Db and Pa proteins and derived the amino acid sequences. Pa was of the same length (150 amino acids) as PRP1 and PIFs but Db had an insert of 21 amino acids at position 82, giving a total of 171 amino acid residues. PIFs has leucine at position 26 instead of the isoleucine in Pa and Db; Pa has cysteine instead of arginine at position 103. This last difference, together with the occurrence of Pa as a dimer linked by an S-S bond between these cysteines at 103, probably explains why Pa is not cleaved and appears as a single band in electrophoresis. The extra insert in Db means that splitting occurs at residue 127 instead of 106 as in the other proteins [6].

The basic proline-rich proteins have been less studied but structures have been worked out or assigned for IB-1, IB-6 (fig. 4) and IB-9 by Kauffman et al. [27–29] and for P-A, P-B, P-C, P-D, P-E, P-F, and P-H by Isemura et al. [30–32] and Saitoh et al. [33–35] and amino acid compositions have been determined for Ps1 and Ps2 [38] and Pc1 and Pc2 [42].

Interrelationships within the basic proline-rich protein group, and between it and the acidic group, are complex and are made even more difficult by the number of proteins described and the different nomenclature used by different groups. It is probably best, therefore, to summarize what is known about each dealing with each classification in turn. IB-1 has 96 residues, including an N-terminal with the first 10 amino acids similar to that of the acidic protein PRP1 and including a phosphoserine at position 8 [4]. Both Pe [40] and Pc [41] proteins are similar to but apparently not identical with IB-1. Kauffman et al. (quoted by Minaguchi and Bennick [6]) have characterized a protein termed DEAE II-2 which has the same N-terminal as IB-1.

IB-6 has 118 residues of which the first 54 are identical with those of IB-9 and the 56 residue C-terminal is identical with P-H [35] and PmS [36]. IB-9 has 54 amino acid residues [27] and is identical with P-E [32] and PmF [36]. In the P series of Isemura et al. [32] and Saitoh et al. [33–35], the structure relationships are as follows: P-A has 38 amino acid residues which correspond with the C-terminal of the 57 residue P-B. P-E has already been described as IB-9 above. P-F has similarities with IB-9 and parts of IB-1 and IB-6. P-H corresponds with the C-terminal 56 residues of IB-6. It is clear that most of these proteins are small and share structural similarities with each other.

The glycosylated proline-rich proteins have not been so closely characterized but the carbohydrate groups are thought to be attached via glucosamine to an asparagine residue [55, 56]. Shibata et al. [57] assigned a structure to the carbohydrate chain, which included glucosamine, fucose, mannose and galactose. The peptide conformation around the attachment site has been modeled by Loomis et al. [58] as a type II′ beta turn.

Despite the large amount of information available on the primary structures of the proline-rich proteins, relatively little progress has been made over their secondary structure. Without the repeating glycine at every third position in collagen, these proteins would not be expected to have a triple helix type of structure despite the high proline content. The repeat sequences suggest folding of the peptide chains [4] but no evidence of these has emerged. Shibata et al. [59] reported a proline helix in a peptide from the basic protein P-D. Cid et al. [60] carried out a computer analysis of possible structures for the acidic proline-rich proteins PRP1 and PRP3 and the basic proline-rich proteins P-D and P-E. PRP1 and PRP3 were predicted to have globular heads with tails of type 3(10) polyproline helices, or to have head and tail in one globular domain due to the tail folding upon itself at positions where random coils were present. P-D and P-E were deduced to be fibrillar proteins with helical zones of the 3(10) type. Braunlin et al. [61], however, from NMR studies of calcium binding by PRP3 concluded that there was extensive motion of the protein residues, conferring a high degree of flexibility on the bound calcium ions. Levine et al. [17] studied a 38,900 dalton glycosylated proline-rich protein from parotid saliva and described the secondary structure as comprising 70% random coil regions and 30% beta turns – the glycosylated domains. Loomis et al. [58] examined the heptapeptide sequence around the glycosylation sites in a proline-rich glycoprotein from parotid saliva. They concluded that a type 11′ beta turn was the most likely configuration although a type I beta turn and a gamma turn were also considered.

The calcium-binding sites of the acidic proline-rich proteins are localized in the N-terminal sequence. Wong and Bennick [46] considered possible similarities with troponin C family of proteins, but realized that the proline content of the binding sequence would change the conformation and weaken the binding of calcium. NMR studies in relation to calcium binding [Madapallimattam and Bennick, referred to in 4] indicated that there were interactions between the 30-residue N-terminal section and the next part of the molecule in PRP3 and PRP1 but that neither histidine nor arginine affected the conformation in this area and arginine was not necessary to maintain folding of the protein.

Two biochemical reactions important in the synthesis of proline-rich proteins have been investigated: the proteolytic cleavage of the precursor proteins at arginine-glycine or arginine-serine linkages already mentioned, and the phosphorylation of the serine residues. The first of these was described by Wong et al. [54] in relation to the production of PRP3 from PRP1 and the salivary enzyme kallikrein implicated as the possible protease involved. Madapallimattam and Bennick [62] found a protein kinase in human parotid salivary glands which was capable of phosphorylating acid proline-rich proteins and other salivary phosphoproteins. It was similar to a monkey parotid gland protein kinase, being activated by magnesium ions, and having a pH optimum between 7.0 and 7.5. Phosphorylation occurred at the serine groups. Activity of the kinase was independent of cAMP, calmodulin, phospholipid or calcium ions. More than one kinase may be necessary.

## *Genetics of Proline-Rich Proteins*

The close similarity in structure of the proline-rich proteins suggests that they may be of similar genetic origin, and indeed attention was early focused on inheritance patterns among this group of proteins. Thus, Azen [63], Azen and Oppenheim [64] and Azen and Denniston [23] were able to identify genetic variation in the proline-rich proteins very soon after Bennick and Connell [19] had isolated the first two. By 1985, population studies and family studies had led to the view that there were at least 13 different loci involved in inheritance of the proline-rich protein types and that in many of these a gene pair directed either synthesis or nonsynthesis of a given protein. This is shown in table 2. The acidic proline-rich proteins were thought to be coded by five separate loci. The first of these, Pr [64], had three alleles, Pr1 coding for PRP1 and PRP3, Pr2 coding for PRP2 and PRP4, and the more

*Table 2.* Gene loci and proline-rich protein products as envisaged up to 1985

| Locus | Alleles | Proteins |
|---|---|---|
| *Acidic* | | |
| Pr | Pr1, Prt2, Pr3 | PRP1, PRP2, PRP3, PRP4 |
| Pa | Pa0, Pa1, Pa2 | Pa1, Pa2 |
| Db | $Db^+$, $DB^-$ | Db f and s |
| PIF | $PIF^+$, $PIF^-$ | PIF f and s |
| *Basic* | | |
| As | $As^+$, $As^-$ | As |
| PmF | $PmF^+$, $PmFF^-$ | PmF (IB-9, P-E) |
| PmS | $PmS^+$, $PmS^-$ | PmS (IB-6) |
| Ps | Ps0, Ps1, Ps2 | Ps1, Ps2 |
| *Basic and glycosylated* | | |
| GI | GI-0, GI-1 to GI-6 | G1 to G6 |
| CON1 | $CON1^+$, $CON1^-$ | CON1 |
| CON2 | $CON2^+$, $CON2^-$ | CON2 |
| Pe | $Pe^+$, $Pe^-$ | Pe |
| Pc | $Pc^+$, $Pc^-$ | Pc |
| Po | $Po^+$, $Po^-$ | Po |
| Pmo1 | $Pmo1^+$, $Pmo1^-$ | Pmo1 |
| Pmo2 | $Pmo2^+$, $Pmo2^-$ | Pmo2 |
| Pe | $P3^+$, $P3^-$ | P3 |

unusual Pr′ coding for PRP3 and a variant of PRP1. Pa [24] also had three alleles, two coding for a protein similar to PRP3 and the other not being expressed. Db [23] was thought to have an allele expressing the protein pair and another being a null allele. Similarly, PIF [25] and the later described As [26] were thought to have expressed and null alleles.

The basic proline-rich proteins were associated with some 9 loci, all with a possible null allele, and only Ps coding for more than one protein. Minaguchi [reported in 6] found five variants of Ps in Japanese populations although Azen and Denniston in black and white Americans had found only two [37]. The other loci are PmF and PmS [37], Pe and Po [40], Pc [41], and three later reported in Japanese populations, Pmo1, Pmo2 [42] and P3 [26].

The glycosylated proline-rich proteins were thought to be coded by three loci, G1 with six or possibly eight alleles [43, 44], and CON1 and CON2 each with an expressed and a null allele [45].

Gene frequencies in different populations are shown in table 3. The total fractional distributions add up to 1 for each locus since the allele was either expressed or not in each case.

*Table 3.* Frequency of occurrence of individual proline-rich proteins in subjects of white, black or Chinese racial characteristics [combined data from 6]

| Protein | Whites | Blacks | Chinese |
|---|---|---|---|
| PRP1/PRP3 | 0.70 | 0.67 | 0.775 |
| PRP2/PRP4 | 0.30 | 0.31 | 0.225 |
| Db | 0.16 | 0.55 | 0.135 |
| No Db | 0.84 | 0.45 | 0.865 |
| PIF | 0.65 | 0.65 | 0.68 |
| No PIF | 0.35 | 0.35 | 0.32 |
| Pa1/Pa2 | 0.15 | 0.16 | 0.20 |
| Pa0 | 0.85 | 0.84 | 0.80 |

## *Early Cloning Studies*

In 1984, Azen et al. [66] reported the cloning of human genes for the salivary proline-rich proteins. The work began with the preparation of a rat copy DNA (cDNA). Treatment of rats with isoproterenol, a beta-adrenergic agonist, stimulates an increase in size and weight of the parotid glands and an increase in the output of proline-rich proteins. Messenger RNA from gland homogenates was then used to prepare cDNA and plasmid clones produced. These clones contained coding regions for repeats of 18 or 19 amino acids. One clone, designated pRP33, contained a cDNA copy of mRNA coding for a proline-rich protein of 186 amino acid residues. A human genomic library was prepared by partially digesting human DNA with the restriction enzyme EcoR1 (from *Escherichia coli*) and incorporating the DNA fragments into the bacteriophage lambda Charon 4A. This Charon 4A library was then screened using $^{32}$P-labeled pRP33 as a probe. The process isolated two clones PRP1 and PRP2 containing EcoR1 fragments of 3.6 and 6.2 kilobase pairs. These were then further studied with other restriction enzymes and the nucleotide sequences identified. In order to shed light on the number of loci coding for proline-rich proteins which appeared to have null alleles, DNA digests were prepared from three parotid glands and one submandibular gland from 4 subjects. No tissue-specific differences were found. No obvious correlations in these and in a further 13 subjects were found between the DNA patterns and the proline-rich protein phenotypes. There was no evidence supporting the possibility of null alleles. Small differences in DNA were found suggesting that deletions or insertions may occur in the gene – this is hardly surprising in view of the repetitive sequences in these genes and their products. It should be commented that it was unfortunate that the Azen group should have decided to give their clones the names PRP1 and PRP2 when these names were already being used for two of the acid proline-rich proteins by other workers and added confusion to an already complex situation.

### *Linkage Studies*

Two types of linkage study have been carried out: those in which linkages between the putative gene loci for the different proline-rich proteins were sought, and those in which linkages between the loci for proline-rich proteins and other gene products were sought in order to localize the proline-rich protein genes to one or more chromosomes [25, 37, 39, 40, 43, 45, 67].

Linkages were demonstrated between Pa-Pr, Pa-Db and Db-G1, Ps-G1 in 299 randomly chosen individuals by Goodman et al. [39]. confirming previous reports (fig. 5). The linkage analysis was carried out by a computer program, LIPED, which generates lod scores (logarithms of the likelihood ratios of multiple linkage, 3 point versus nonlinkage of loci) and the maximum estimate of the recombination fraction, theta max. Higher lod scores indicate linkage. Examination of the values obtained showed that the eight previously described loci Pa, Pr, Db, G1, Ps, PIF, PmF, PmS had scores of 3 or higher with at least one of the other loci These results therefore gave strong indications that the salivary protein genes were closely linked. The term salivary protein complex (SPC) was coined to describe this complex of genes. All the proteins coded were similar in general structure.

Analysis of the linkages between the SPC and other markers led Goodman et al. [39] to conclude that although some scores had suggested the short arm of chromosome 6 as a possible site for the SPC, there was also considerable evidence against this site.

A further paper from the same group in 1985 [66] approached the problem in a different way. Using the PRP1 probe described above, they prepared a 980 base pair fragment from the coding region with the restriction enzyme HinfI. Hybrid cell lines were prepared by fusing human cells with rodent cell lines. The DNA from these hybrids was probed with the proline-rich protein coding probe and the proline-rich protein gene was found to be associated with chromosome 12. This association was confirmed using a probe for a 12q restriction fragment. It is now established, therefore, that the genes coding for the proline-rich proteins are situated as a complex on the

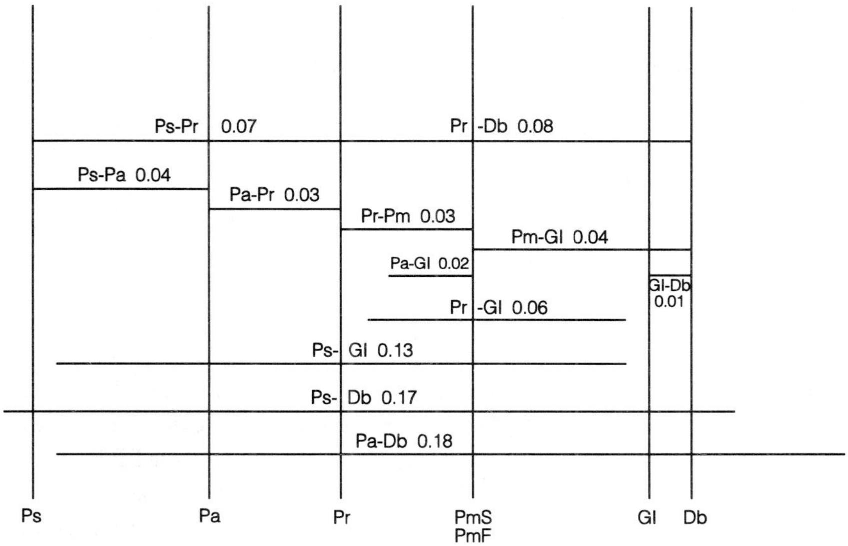


*Fig. 5.* Calculated distances between loci coding for proline-rich proteins within the SPC. Data from Goodman et al. [39]. The vertical lines indicate the most probable spacing derived from the individual distances; the individual calculated distances are shown by appropriately labeled horizontal lines.

short arm of chromosome 12. Goodman et al. [39] estimate that the SPC is about 15 cm long.

### *Further Studies of Inheritance*

Up to this point the SPC was still seen as involving at least 13 loci with null alleles for most of the phenotypes. Maeda [69], in a paper which presented little new data, but which presented a remarkable re-interpretation of existing and unpublished data, produced a simplifying hypothesis which reduced the number of loci to six, arranged in subfamilies. The DNA clones prepared by Azen et al. [66], PRP1 and PRP2, had had their partial nucleotide sequences described and these matched with the decoded sequences for amino acids in the proline-rich protein repeat sequences. The probe from PRP1 used to probe an EcoR1 digest of total human DNA had yielded only six hybridizing fragments (including two matching to PRP1 and

*Table 4.* Frequency of occurrence of proline-rich proteins in different populations re-examined on a basis of Maeda's scheme of inheritance (data as table 2)

| Protein | Whites | Blacks | Chinese |
|---|---|---|---|
| Db | 0.16 | 0.55 | 0.135 |
| IF | 0.65 | 0.35 | 0.68 |
| Pa1/Pa2 | 0.15 | 0.16 | 0.20 |
| Total | 0.96 | 1.06 | 1.015 |

PRP2). The polymorphisms of these fragments did not match the genotypes deduced from the observed phenotypes. There was no evidence of weak or absent bands which might be associated with null alleles for PRP1 or PRP. All six EcoR1 fragments were found to contain only one gene sequence. Further, cloning experiments [70] (to be described in detail in the next section) produced only two types of clone – one with repeated nucleotide sequences which could be split by the restriction enzyme BstN1 (already described by Azen et al. [66]), and the other with a slightly different repeat which could be split by another restriction enzyme, HaeIII. Probes made from the repeat sequences identify six EcoR1 fragments from human genome DNA – four EcoR1 fragments hybridize with the BstN1 repeat probes and two with the HaeIII repeat probes. These observations led Maeda to re-examine previous structural and population **data.** Four of the acidic proline-rich proteins show double-banded phenotypes – PRP1 and PRP3, PRP2 and PRP4, PIF slow and fast, Db slow and fast. The PRP3 and PRP4 proteins are thought to arise from their partners by proteolysis at an arginine group (106 in the case of PRP1) and it is therefore probable that the arginine at position 6 in the PRPG and PIF doublets, and the equivalent arginine at position 127 in the Db doublet, is the point at which a split can occur to generate the shorter member of the pair from the larger. The fifth major acidic proline-rich protein Pa does not exhibit a double band. It is a disulphide-bonded dimer with a cysteine group at 106 instead of arginine [53]. If Pa, Db and PIF were all coded at the same locus, then there should be no subjects who could synthesize all three proteins, and yet no-one who did not synthesize at least one of them. Previous data from Azen [71] and Azen and Denniston [25] confirmed this (table 4). Further, the sum of the gene frequencies for Db, Pa1, Pa2 and PIF was approximately one in the populations studied and the data from family studies were consistent with the hypothesis. Thus one locus

*Table 5.* Gene nomenclature as proposed by Maeda [69]

| Type as determined by restriction enzymes | Locus | Products |
|---|---|---|
| *Acidic PRPs* | | |
| HAEIII | PRH1 | Db, Pa1, Pa2, PIF |
| | PRH2 | PRP1, PRP3 |
| *Basic and glycosylated PRPs* | | |
| BstN1 | PRB1 | |
| | PRB2 | |
| | PRB3 | G1- |
| | PRB4 | |

might code for the proteins PRP1, PRP2, PRP3 and PRP4, and a second for the Db, Pa and PIF groups. Maeda therefore put forward the scheme shown in table 5.

Later work has supported this analysis and the gene frequencies given by Minaguchi and Bennick [6] can be fitted into the hypothesis. The variety of proline-rich proteins arises in part from post-translational processing.

### *Further Cloning Studies*

After the first cloning studies on the genes for proline-rich proteins by Azen et al. [47] described above, there followed more extensive work, notably by Maeda et al. [70], Mamula et al. [72] and Azen et al. [53]. Maeda et al. [70] prepared mRNA from a parotid gland removed from a patient in the course of cancer surgery. The patient's saliva contained acidic proline-rich proteins PRP1, PRP3, and PIF, and the basic proline-rich proteins Ps1, and the glycosylated proline-rich proteins G1-1, G1-2 and G1-3. Db, Ps1, PmF (IB-9) and PmS (IB-6) were absent. Several clones were produced, two coding for proline-rich proteins which contained repeat sequences split by HaeIII and five coding for basic or glycosylated proline-rich proteins which had repeat sequences which could be split by BstN1. All the clones had the same signal region and N-terminal sequence although an additional N-terminal sequence was present in the two clones coding for acidic proline-rich proteins. Thus these two clones appeared to correspond to the PRH1 and PRH2 loci suggested by Maeda. Three of the clones coding for repeats splittable by BstIII correspond with PRB1 locus, and one was very similar to the clone PRP1 produced earlier by Azen et al. [47]. Another of the new clones was similar to the clone PRP2 of Azen et al. [47] and was assigned to the PRB2 locus. The fifth clone was probably assignable to the PRB4 locus.

Later, in 1987, Azen et al. [53] succeeded in cloning the entire exon and intron structures of the genes for Db and Pa. Mamula et al. [72] with yet another clone were able to demonstrate proline-rich protein mRNA production actually in the acinar cells of the parotid gland.

## *The Present State of Knowledge*

Several important conclusions emerge from these studies. Firstly, the majority of the proteins of saliva are of similar structural composition and contain similar repeat sequences of amino acids. This, together with the small differences which result in overall acidic or basic charge and the possibility of glycosylation, permits a spectrum of activities which could enhance their protective functions in the mouth. Secondly, despite the variety of proteins present, they are coded by a very small number of repeat sequences of DNA. One predominant repeat sequence in the clones resulting from HaeIII cleavages codes for the acidic proline-rich protein structures and some four sequences in the clones resulting from BstN1 cleavages give the basic and glycosylated proteins. Differential splicing of the RNA at the many repeat sections increases the number of possible proteins originating from a small number of genes, and this number is further increased by the possibility of post-translational cleavage by enzymes such as kallikrein. Thirdly, the production of the proline-rich proteins appears to be controlled by a very small group of genes, probably six in number, which are situated in the SPC on human chromosome 12. Since closely similar genes are found in animals other than man, the proline-rich proteins would appear to be an evolutionarily stable group of molecules.

## *References*


1 Mandel ID, Thompson RH, Ellison SA: Studies on the mucoproteins of human parotid saliva. Arch Oral Biol 1965;10:499–507.

2 Bennick A: Chemical and physical characterisation of a phosphopeptide, protein C, from human saliva and comparison with a related protein A. Biochem J 1977; 163:229–239.

3 Azen EA: Genetic protein polymorphisms of human saliva; in Tenovuo JO (ed): Human Saliva: Clinical Chemistry and Microbiology. Boca Raton, CRC Press, 1989, vol 1, pp 161–195.
4 Bennick A: Structural and genetic aspects of proline-rich proteins. J Dent Res 1987; 66:457–461.
5 Hay DI, Moreno EC: Statherin and the acidic proline-rich proteins; in Tenovuo JO (ed): Human Saliva: Clinical Chemistry and Microbiology. Boca Raton, CRC Press, 1989, vol 1, pp 131–150.
6 Minaguchi K, Bennick A: Genetics of human salivary proteins. J Dent Res 1989;68: 2–15.
7 Kauffman DL, Keller P: The basic proline-rich proteins in human parotid saliva from a single subject. Arch Oral Biol 1979;24:249–256.
8 Schlesinger DH, Hay DI: Complete covalent structure of a proline-rich phosphoprotein, PRP2, an inhibitor of calcium phosphate crystal growth from human parotid saliva. Int J Pept Protein Res 1986;27:373–379.
9 Warner TF, Azen EA: Proline-rich proteins are present in serous cells of submucosal glands in the respiratory tract. Am Rev Resp Dis 1984;130:115–118.
10 Ito S, Isemura S, Saitoh E, Suzuki T, Sanada K, Shibata A: A study on the salivary proline-rich peptide P-C with special reference to the presence of salivary peptide P-C in the human respiratory tract. Folia Endocrinol Jap 1984;60:16–22.
11 Ito S, Isemura S, Saitoh E, Sanada K, Suzuki T, Shibata A: Immunohistochemical demonstration of salivary proline-rich peptide P-C-like immunoreactivity in human pancreatic beta-cells. Acta Endocrinol 1983;103:544–551.
12 Hagerman AE, Butler LG: The specificity of proanthocyanidine-protein interactions. J Biol Chem 1981;256:4494–4497.
13 Bennick A, McLaughlin AC, Grey AA, Madapallimattam G: The location and nature of the calcium-binding sites in salivary acidic proline-rich proteins. J Biol Chem 1981;256:4741–4746.
14 Moreno EC, Varughese K, Hay DI: Effect of human salivary proteins on the precipitation kinetics of calcium phosphate. Calcif Tissue Int 1979;28:7–16.
15 Kousvelari EE, Baratz RS, Burke B, Oppenheim FG: Immunological identification and determination of proline-rich proteins in salivary secretion, enamel pellicle and glandular tissue. J Dent Res 1980;59:1430–1438.
16 Gibbons RJ, Hay DI: Human salivary acidic proline-rich proteins and statherin promote the attachment of *Actinomyces viscosus* LY7 to apatitic surfaces. Infect Immun 1988;56:439–445.
17 Levine MJ, Reddy MS, Tabak LA, Loomis RE, Bergey EJ, Jones PC, Cohen RE, Stinson MW, Al-Hashimi I: Structural aspects of salivary glycoproteins. J Dent Res 1987;66:436–441.
18 Yu PL, Bixler D, Goodman PA, Azen EA, Karn RC: Human parotid proline-rich proteins: correlation of genetic polymorphisms to dental caries. Genet Epidemiol 1986;3:147–152.
19 Bennick A, Connell GI: Purification and partial characterisation of four proteins from human parotid saliva. Biochem J 1971;145:557–567.
20 Oppenheim FG, Hay DI, Franzblau C: Proline-rich proteins from human parotid saliva. I. Isolation and partial characterisation. Biochemistry 1971;10:4233–4238.
21 Hay DI, Oppenheim FG: The isolation from human parotid saliva of a further group of proline-rich proteins. Arch Oral Biol 1974;19:627–632.

22 Bennick A: Structure and function of salivary proline-rich proteins. Mol Cell Biochem 1982;45:83–99.
23 Azen EA, Denniston CL: Genetic polymorphism of human salivary proline-rich proteins: further genetic analysis. Biochem Genet 1974;12:109–120.
24 Friedman RD, Merritt AD: Partial purification and characterisation of a polymorphic protein (Pa) in human parotid saliva. Am J Hum Genet 1975;27:304–314.
25 Azen EA, Denniston CL: Genetic polymorphism of PIF (parotid isoelectric focusing variant) proteins with linkage to the PPP (parotid proline-rich protein) gene complex. Biochem Genet 1981;19:475–485.
26 Minaguchi K, Suzuki K: New salivary protein polymorphisms detected by SDS polyacrylamide gel electrophoresis. J Dent Res 1986;65:326.
27 Kauffman DL, Wong RSC, Bennick A, Keller PJ: Basic proline-rich proteins from human parotid saliva: complete covalent structure of protein IB-9 and partial structure of protein IB-6, members of a polymorphic pair. Biochemistry 1982;21: 6558–6562.
28 Kauffman DL, Hoffman T, Bennick A, Keller P: Basic proline-rich proteins from human parotid saliva: complete covalent structure of proteins IB-1 and IB-6. Biochemistry 1986;25:2387–2392.
29 Kauffman DL, Bennick A, Blum M, Keller P: Basic proline-rich proteins: sequences and relationship to cDNA sequences. J Dent Res 1989;68:1024.
30 Isemura S, Saitoh E, Sanada K: Isolation and amino acid sequences of proline-rich peptides of human whole saliva. J Biochem 1979;86:79–86.
31 Isemura S, Saitoh E, Sanada K: The amino acid sequence of a salivary proline-rich phosphoprotein. J Biochem 1980;87:1071–1077.
32 Isemura S, Saitoh E, Sanada K: Fractionation and characterisation of basic proline-rich peptides of human parotid saliva and the amino acid sequence of a proline-rich peptide P-E. J Biochem 1982;91:2067–2075.
33 Saitoh E, Isemura S, Sanada K: Complete amino acid sequence of a basic proline-rich peptide, P-D, from human parotid saliva. J Biochem 1983;93:495–502.
34 Saitoh E, Isemura S, Sanada K: Complete amino acid sequence of a basic proline-rich peptide, P-F, from human parotid saliva. J Biochem 1983;93:883–888.
35 Saitoh E, Isemura S, Sanada K: Further fractionation of basic proline-rich peptides from human parotid saliva and complete amino acid sequence of basic proline-rich peptide P-H. J Biochem 1983;94:1991–1999.
36 Anderson LC, Kauffman DL, Keller P: Identification of the Pm and PmS human parotid salivary proteins as basic proline-rich proteins. Biochem Genet 1982;20:1131–1137.
37 Azen EA, Denniston CL: Polymorphism of Ps (parotid size variant) and detection of a protein (PmS) related to the Pm (parotid middle band) system with genetic linkage of Ps and Pm to G1, Db and Pr genetic determinants. Biochem Genet 1980;18:483–501.
38 Goodman PA, Karn RC: Human parotid size polymorphism (Ps): characterisation of two allelic products, Ps 1 and 2, by limited proteolysis. Biochem Genet 1983;21: 405–416.
39 Goodman PA, Yu PL, Azen EA, Karn RC: The human salivary protein complex (SPC): a large block of related genes. Am J Hum Genet 1985;37:785–797.
40 Azen EA, Yu PL: Genetic polymorphism of Pe and Po salivary proteins with probable linkage of their genes to the salivary protein gene complex (SPC). Biochem Genet 1984;22:1065–1080.

41 Karn RC, Goodman PA, Yu PL: Description of a genetic polymorphism of a human proline-rich salivary protein, Pc, and its relationship to other proteins in the salivary protein complex (SPC). Biochem Genet 1985;232:37–51.
42 Minaguchi K, Suzuki K: Genetic polymorphism of the Pmo1 and Pmo2 salivary proteins detected by the modified protein staining method. Ann Hum Genet 1988; 52:177–187.
43 Azen EA, Hurley CK, Denniston C: Genetic polymorphism of the major parotid salivary glycoprotein (G1) with linkages to the genes for Pr, Db, and Pa. Biochem Genet 1979;17:257–279.
44 Minaguchi K, Takeusu Y, Tsutsumi T, Suzuki K: Studies of genetic markers in human saliva. VII. Frequencies of the major parotid salivary glycoprotein (G1) system in a Japanese population. Bull Tokyo Dent Coll 1981;22:1–6.
45 Azen EA, Yu PL: Genetic polymorphism of CON1 and CON2 salivary proteins detected by immunologic and concanavalin A reactions on nitrocellulose with linkage of CON1 and CON2 genes to the SPC (salivary protein gene complex). Biochem Genet 1984;22:1–19.
46 Wong RSC, Bennick A: The primary structure of a salivary calcium-binding proline-rich phosphoprotein (protein C), a possible precursor of a salivary-related protein A. J Biol Chem 1980;255:5943–5948.
47 Azen EA, Lyons KM, McGonigal T, Barrett NL, Clements LS, Maeda N, Uanin EF, Carlson DM, Smithies O: Clones for the human gene complex coding for salivary proline-rich proteins. Proc Natl Acad Sci USA 1984;81:5561–5565.
48 Wong RSC, Hofmann T, Bennick A: The complete primary structure of a proline-rich phosphoprotein from human saliva. J Biol Chem 1979;254:4800–4808.
49 Friedman RD, Karn RC: An hypothetic mechanism concerning the origin of proline-rich human salivary proteins. Am J Hum Genet 1977;29:44A.
50 Karn RC, Friedman RD, Merritt AD: Human salivary proline-rich (Pr) proteins: A post-translational derivation of the phenotypes. Biochem Genet 1979;17:1061–1077.
51 Schlesinger DH, Hay DI: Complete primary structure of a proline-rich phosphoprotein (PRP-4), a potent inhibitor of calcium phosphate precipitation in human parotid saliva; in Gross E, Meienhofer J (eds): Peptides. Proc 6th American Peptide Symposium, Rockford, Pierce Chemical Co, 1979, pp 133–136.
52 Hay DI, Bennick A, Schlesinger DH, Minaguchi K, Madapallimattam G, Schluckebier SK: The primary structures of six human salivary acidic proline-rich rich proteins (PRP-1, PRP-2, PRP-3, PRP-4, PIF-s and PIF-f). Biochem J 1988;255:15–21.
53 Azen EA, Kim HS, Goodman P, Flynn S, Maeda N: Alleles at the PRH1 locus coding for the human salivary acidic proline-rich proteins Pa, Db and PIF. Am J Hum Genet 1987;41:1035–1047.
54 Wong RSC, Madapallimattam G, Bennick A: The role of glandular kallikrein in the formation of a salivary proline-rich protein A by cleavage of a single bond in salivary protein C. Biochem J 1983;211:35–44.
55 Degand P, Boersma A, Roussel P, Richet C, Biserte G: The peptide moiety of the region carrying the carbohydrate chains of the human parotid glycoprotein. FEBS Lett 1975;54:189–192.
56 Li HC, Levine MJ: Characterisation of a glycopeptide from the proline-rich glycoprotein of human parotid saliva. Arch Oral Biol 1980;25:353–355.

57 Shibata S, Nagata K, Nakamura R, Tsunemitsu A, Misaki A: Interaction of a parotid saliva basic glycopeptide with *Streptococcus sanguis* ATCC 10557. J Periodontol 1980;51:499–504.
58 Loomis RE, Prakobphol A, Levine MJ, Reddy MS, Jones PC: Biochemical and biophysical comparison of two mucins from human submandibular-sublingual saliva. Arch Biochem Biophys 1987;258:452–464.
59 Shibata S, Asakura J, Isemura T, Isemura S, Saitoh E, Sanada K: Conformational study of basic proline-rich polypeptides from human parotid saliva. Int J Pept Protein Res 1984;23:158–165.
60 Cid H, Uargas U, Bunster M, Bustos S: Secondary structure prediction of human salivary proline-rich proteins. FEBS Lett 1986;198:140–144.
61 Braunlin WH, Vogel HJ, Drakenberg T, Bennick A: A calcium-43 study of calcium binding to an acidic proline-rich phosphoprotein from human saliva. Biochemistry 1986;25:584–589.
62 Madapallimattam G, Bennick A: Phosphorylation of salivary proteins by a salivary gland protein kinase. J Dent Res 1986;65:405–411.
63 Azen EA: Genetic polymorphism of basic proteins from parotid saliva. Science 1972; 176:673–674.
64 Azen EA, Oppenheim FG: Genetic polymorphism of proline-rich human salivary proteins. Science 1973;180:1067–1069.
65 Friedman RD, Merritt AD, Rivas ML: Genetic studies of human acidic salivary protein (Pa). Am J Hum Genet 1975;27:292–303.
66 Azen EA, Goodman PA, Lalley PA: Human salivary proline-rich protein genes on chromosome 12. Am J Hum Genet 1985;37:418–424.
67 Yu PL, Karn RC, Merritt AD, Azen EA, Conneally PM: Linkage relationships and multipoint mapping of human parotid salivary proteins (Pr, Pa, Db). Am J Hum Genet 1980;32:555–563.
68 Mamula PW, Heerema NA, Palmer CG, Lyons KM, Karn RC: Localisation of the human salivary protein complex (SPC) to chromosome band 12p13.2. Cytogenet Cell Genet 1985;39:279–284.
69 Maeda N: Inheritance of the human salivary proline-rich proteins: a reinterpretation in terms of six loci forming two subfamilies. Biochem Genet 1985;23:455–464.
70 Maeda N, Kim HS, Azen EA, Smithies O: Differential RNA splicing and post-translational cleavages in the human salivary proline-rich protein gene system. J Biol Chem 1985;260:11123–11130.
71 Azen EA: Genetic protein polymorphism in human saliva: an interpretive review. Biochem Genet 1978;16:79–99.
72 Mamula PW, Morley DJ, Larsen SH, Karn RC: Expression of human salivary protein genes. Biochem Genet 1988;26:165–175.
73 Robinson R, Kauffman DL, Waye MMY, Blum M, Keller PJ, Bennick A: Differential expression of genes for proline-rich proteins in salivary glands. J Dent Res 1989:68:1024.

D.B. Ferguson, PhD, BDS, Department of Physiological Sciences,
University of Manchester, GB-Manchester M13 9PT (UK)

Ferguson DB (ed): Aspects of Oral Molecular Biology.
Front Oral Physiol. Basel, Karger, 1991, vol 8, pp 117–140

# Genetic Control of Mammalian Salivary Proteins[1]

*Robert C. Karn*

Department of Biological Sciences, Butler University, Indianapolis, Ind., USA

## *Proteins in the Salivas of Mammals*

Mammalian salivas contain a variety of proteins, some with relatively well-defined functions and others whose functions remain to be elucidated. Many of these proteins are unusual in their structures and/or properties. Most of the studies of mammalian salivary proteins have been done in relatively few organisms. These include humans, some species of monkey, laboratory rats and mice. The majority of salivary proteins are the secretory products of one or more of three different salivary glands: the parotid, the sublingual and the submandibular. Thus it is the expression of the genes in these three tissues that is the concern of this chapter.

Salivary secretions vary because of several factors. The volume of fluid secreted per unit time from any of the three types of glands influences total protein concentration. The rates of synthesis and degradation of secreted proteins affect both the concentration of total protein and the composition of proteins in saliva. This chapter will not deal with the physiological factors affecting protein concentration but will cite instances in which gene regulation has been shown to affect the composition of proteins in saliva.

[1] This chapter is dedicated to the memories of A. Donald Merritt and Pao-Lo Yu, two investigators who contributed much to our understanding of human salivary protein genetics.

## *Characterization of Salivary Proteins*

Some salivary proteins are enzymes. The best characterized of these is salivary amylase, a major contribution to the salivas of humans and rodents [1, 2]. Other enzymes have been identified but have received less attention. It is known that human saliva contains peroxidase [3] and phosphohydrolases [4] but these appear in much lower quantity than amylase. Kallikrein [5] is also produced in the parotid gland and is important in processing some salivary proteins as will be discussed later.

Many salivary proteins have been described in terms of their unusual amino acid composition rather than their specific function. Among these are the proline-rich proteins (PRP) which are the largest single group in human saliva [6–9]. They constitute about 70–80% of human salivary proteins. The PRP were originally identified by the predominance of proline, glycine and glutamic acid/glutamine in their amino acid compositions. Conversely, they contain no cysteine (with one exception among the human PRP) and no methionine. The basic subgroup contains no aromatic amino acids and no branched-chain amino acids. Other groups of human salivary proteins identified by their unusual amino acid compositions are the tyrosine-rich [10], the histidine-rich (histatins) [11] and the cysteine-containing (cystatins) [12] phosphoproteins. Parotid secretory protein (PSP) in mouse saliva has been characterized as rich in leucine [13].

A third means of characterizing salivary proteins has been observation of the ability to bind one or more ligands. Grasbeck et al. [14] isolated a protein identified by its ability to bind vitamin $B_{12}$ from human saliva and characterized it. Although these properties were not the original means by which they were characterized, the PRP have been reported to bind calcium [7, 8] and tannins [15]. More recently, a group of mouse salivary proteins, which have been shown to bind androgens, have been described and characterized biochemically and genetically [16].

Other proteins are found in trace amounts in human saliva, including serum proteins or their fragments, enzymes not mentioned above and hormones. A discussion of these minor components is beyond the scope of this chapter.

## *Functions of Salivary Proteins*

The variety of tasks performed by the proteins in saliva is impressive. The mucins provide protective barriers to the hard and soft tissues of the oral cavity, lubricating and guarding them against damage from mechanical. chemical and microbial sources (see chapter by Tabak, pp. 77–94). The histatins have been observed to have an antimicrobial role in saliva [11, 17], although other salivary proteins may also participate in this function [18]. The cystatins are thiol protease inhibitors [19].

A variety of soluble enzymes, most of which have hydrolytic activities, are found in the mixture. Their functions are defined to some extent by their specificities. For example, salivary amylase has a relatively obvious role in digestion of dietary starch. The actions of others are not entirely clear. The protease kallikrein has been shown to be capable of processing precursors of salivary proteins and peptides, but it may have other functions. The trace amounts of phosphohydrolases, peroxidase and glucose-6-phosphate dehydrogenase have yet to be explained.

The functions of various ligand-binding proteins are also speculative. The PRP have been implicated in maintaining the supersaturated level of calcium in saliva and preventing formation of hydroxyapatite crystals on tooth surfaces [7, 8]. Their ability to bind to hydroxyapatite causes them to participate in the formation of dental pellicle [7, 8]. Carlson and co-workers [20–22] have also suggested that these proteins are responsible for binding dietary tannins in rodents to neutralize the potentially harmful effects of these polyphenols to the organism.

Close linkage, without observed recombinations, of mouse PRP to the bitter taste genes *Qui* (quinone), *Rua* (raffinose acetate) and *Cyx* (cyclohexamide) led to the suggestion that PRP might be involved in taste [23] but more recent findings of recombination of PRP and the taste genes [24] have made that less likely. A study of human PRP genes and taste for phenythiocarbamide (PTC) suggests that the PRP genes are not linked to PTC taste [25]. Several other ligand-binding proteins in saliva have roles that are presently undefined but which might be involved in taste and/or olfaction. A highly speculative role for the mouse submandibular androgen-binding proteins is oral and/or olfactory recognition that may involve discrimination of the sexes or other species. This is discussed later.

### *Amylase Genetics*

One of the first salivary proteins to attract attention was amylase. Its enzymatic function is the hydrolysis of the alpha-1,4 linkages in starch molecules [1]. The high level of amylase activity in some mammalian salivas makes it easy to detect by traditional assay methods. As will be evident in the discussion below, any treatment of salivary amylase without reference to other mammalian amylases would be an incomplete story. Therefore, a concise picture of the genetic control of mammalian amylases will be presented here in order that salivary amylase genetics are seen in perspective. A number of reviews [1, 2, 26, 27] have treated mammalian amylases in detail, so only the key features of this fascinating genetic system will be described here.

The primary source of amylase in nearly all vertebrates is the pancreas. Salivary amylase is found only in primates, rodents and some species of frog [1]. In humans the parotid gland accounts for over 50% of amylase in mixed saliva where the enzyme accounts for as much as 30% of the protein [28]. To a lesser extent, the salivary-type isozymes are produced in lacrimal duct and breast tissue in humans [2]. There have been reports that a third type of amylase is produced in the ovaries [2] and fallopian tubes [29] but those observations have not been confirmed. Likewise a messenger RNA for amylase nearly identical to that made in the salivary glands has been found in mouse liver at levels substantially below those of the salivary glands and pancreas [30, 31] and there is evidence for low-level expression of the pancreatic amylase gene in human liver [32].

Mammalian amylases are monomeric proteins composed of approximately 500 amino acids. They contain relatively high levels of tyrosine and tryptophan as well as dicarboxylic acids/amides [1]. They have pyroglutamic acid N-termini, probably as the result of the spontaneous cyclization of glutamine which becomes the N-terminal residue when the signal peptide is removed in synthesis [33]. Creation of a pyroglutamic N-terminus upon removal of the signal peptide appears to be a characteristic of a number of exocrine pancreatic proteins [34]. Amylases contain calcium ion which has a role in stabilizing the structure over a broad pH range [35]. In human saliva, a series of isozymes of amylase have been observed as the result of both glycosylation and deamidation [36].

Two groups of genes for amylase have been described: salivary and pancreatic. Observation of close linkage between these genes led to the proposal that modern human amylase genes evolved by duplication of ancestral genes [36]. Genetic studies suggested the same explanation for mouse [37] and vole [38] amylases. These proposals were later supported by sequence data for mouse salivary and pancreatic amylase cDNAs which proved to be 10% different [39]. The salivary cDNA also contained nine bases not found in the pancreatic sequence. Comparison of pancreatic sequences from various mammals suggests that the nine bases were lost during the evolution of rodent pancreatic amylases since the gap does not appear in the pancreatic amylase sequences of other mammals [40].

It is now known that human and rodent amylases are encoded by multigene families [26, 27]. It appears that there is only one salivary gene in most inbred strains of mice which have been examined but there may be two or three in the YBR strain [26]. Hamsters and voles may also have multiple salivary genes [38, 41, 42]. In addition, the mouse salivary amylase gene appears to be responsible for the low level of mRNA in liver alluded to above. An amylase cDNA cloned from the RNA of mouse liver proved to be identical to a mouse salivary amylase cDNA in all respects except for the beginning of the 5′ nontranslated leader sequence [30, 31]. The liver mRNA results from expression of a liver-specific promoter and a nontranslated exon located between the salivary promoter and the transcription initiation site [43]. Recent analyses of amylase expression in mice transgenic for the YBR Amyl.1 gene suggest that expression of that gene in the liver is responsible for the appearance of the YBR-specific salivary isozyme (AMY1B) in the serum of the transgenic mouse [44]. The regulation of expression of amylases is discussed later in this chapter.

Quantitative studies of mouse pancreatic amylase isozymes suggested that pancreatic amylase genes are present in as many as 20 copies [45]. The number of genes has since been demonstrated to vary in different mouse strains [27]. Two different pancreatic isozyme sequences with corresponding cDNAs and genes were demonstrated in strain YBR [46–48] while five different cDNAs have been isolated from strain CE [49]. Quantitative estimates based on hybridization of cDNAs to genomic DNA suggest that there may be as many as twelve gene and pseudogene copies per haploid genome of some mice [reviewed in 27].

Human amylases are also encoded by a multigene family. As early as 1982, Pronk et al. [50] produced genetic evidence for two human salivary amylase genes. It was not until 1984, however, that human amylase cDNA sequences became available, and 1986 before a human amylase gene sequence was published.

The first sequences of human salivary amylase and pancreatic amylase cDNAs were published by Nakamura et al. [51]. In a report that followed shortly after, another group noted significant differences between the pancreatic amylase sequence of Nakamura et al. and that of a cDNA produced from poly(A+)RNA from an individual shown to have been homozygous for the common $Amy_2^A$ allele [40]. Most of the differences were errors in the first pancreatic sequence and corrections of both the Nakamura sequences were later published [52]. The first reported sequence of a human salivary amylase gene described the exons [53] and compared them to the cDNA sequences derived in the same laboratory [51]. The gene they described was about 10 kilobases (kb) in size and they presented the sequences of eleven exons which are separated by ten introns. Shortly thereafter, the sequence of approximately three-quarters of a salivary amylase gene, which includes 1,500 base pairs of flanking sequence, the first six exons, the first five introns and the last exon appeared [54]. That study focused on potential control regions of the gene, identifying a possible translation initiation (CAP) site, a TATA box, a CAAT box and a Z DNA sequence. In addition, a TATA box-like sequence of unknown function was found in the second exon. That report called attention to two general motifs, TGGCT and GGGCT, which are repeated at a higher frequency upstream of the ATG start codon than at any other place in the sequence, proposing that these and other five-base sequences found in the gene could be enhancers.

A study of the human amylase multigene family has identified three salivary and two pancreatic amylase genes along with two pseudogenes [55]. These are purported to represent all the amylase genes in a human genome (see, however, the haplotype discussion below). This has several important consequences. First, experiments comparing these genes with the mouse liver promoter and NTE sequences showed that no equivalent to the sequences for liver-specific expression in the mouse occur in human amylase genes. That agrees with biochemical genetic studies [56] and later molecular studies [53] that suggested amylase genes were not expressed in human liver. More recently, however, the same group has reported finding transcripts of the pancreatic gene at low levels in human liver tissue [32]. Second, an obvious interpretation of their results leads to the conclusion that, if there is amylase expression in female genital tract, it must be that of one or more of the three salivary and two pancreatic genes described. There does not appear to be a unique gene that could account for a third amylase type.

Very recent evidence published by Groot et al. [57] shows that amylase genes exist as at least two haplotypes with significantly different numbers of salivary amylase genes. They describe a short haplotype comprised of two pancreatic genes (AMY2A and AMY2B) and one salivary gene (AMY1C) arranged in the order 2B-2A-1C and with a total length of ca. 100 kb. The alternative is a long haplotype containing six more genes which they propose to be arranged in two repeats, each with two salivary genes (AMY1A and AMY1B) and a pseudogene (AMYP1) lacking the first three exons of salivary amylase genes. This haplotype is ca. 300 kb long with the order 1A-1B-P1 in each repeat. These authors propose a model for evolution of the human alpha-amylase multigene family with the general arrangement of the repeats described above as 2B-2A-$(1A-1B-P1)_n$-1C, where n is either 0 or 2 to account for the short (designated AMY*HO) and long (designated AMY*H2) haplotypes. They cite general models for evolution of multigene families and propose that the two haplotypes may have originated by unequal, homologous recombination between two copies of a haplotype carrying one copy of the direct repeat with the structure AMY2B-AMY2A-(AMY1A-AMY1B-$AMYP1)_1$-AMY1C (designated AMY*H1). The cluster can expand, by unequal crossing over, three genes at a time (i.e. 1A-1B-P1). They point out that a clear distinction should be made in this case between haplotypes characterized at the DNA level as opposed to haplotypes characterized at the protein (i.e. gene expression) level since DNA haplotypes may differ while protein haplotypes do not.

Long before the existence of multigene families was suspected, studies were under way to determine the chromosomal locations of mammalian amylases. Early linkage studies assigned the human salivary and pancreatic amylase genes to an area of chromosome 1 near the centromere [2]. More recently, in situ hybridization studies pinpointed their location to 1p21 [58]. The mouse amylase gene cluster has been assigned to chromosome 3 [59].

### *Proline-Rich Protein Genetics*

The chapter by Ferguson (pp. 95–116) is devoted to the genetics of PRP, so only the highlights of this system will be summarized here. PRP were described as constituents of human saliva nearly 25 years ago [60] and have since been identified in the respiratory tract [61] and the pancreas [62]. As indicated by their name, this group of salivary proteins is rich in proline but also contains high levels of glycine and glutamic acid/glutamine. They represent a variable proportion of salivary proteins, depending on the species and conditions. PRP production is essentially constitutive in humans and at least some nonhuman primates, where these proteins represent as much as 70–80% of the total protein concentration in parotid saliva. In rodents, however, PRP production must be induced, for example by long-term isoproterenol stimulation [63–66] or by inclusion of tannins in the diet [15, 20–22, 65].

Historically, two lines of investigation led to the characterization of human parotid PRP – biochemical and genetic. Based on biochemical comparisons, human PRP have been separated into three general categories: acidic, basic and glycosylated PRP. In addition to being high in proline, glycine and glutamic acid/glutamine, these share the characteristic of being low in nonpolar amino acids.

The 70s and 80s saw the descriptions of as many as 17 different human PRP polymorphisms, found in the literature as Pr [67], Pa [68], Db [69], Gl [70], PmS [71, 72], PmF [71, 72], Ps [72], PIF [73], CON1 [74], CON2 [74], Pe [75], Po [75], Pc [76], and As, Pmo1, Pmo2 and P3 [77]. As these were being described, early attempts at determining linkage, both among the PRP markers and with other human polymorphisms, were under way. Eventually seven of the PRP markers, Pa, Pr, Db, Gl, PmS, PmF and Ps, constituted a gene cluster designated the salivary protein complex (SPC) [78].

Attempts to relate salivary protein markers to other human polymorphisms, with the goal of making a chromosomal assignment, were unsuccessful. Tentative linkages to chromosome 6 markers based on positive linkage scores with properidin factor B (Bf) and glyoxylase (GLO-I) [79] were not supported by further studies. Assignment of the SPC was eventually realized via hybridization of a human genomic PRP probe with the DNAs from mouse-human hybrid cell lines [80, 81]. Their results suggested that the SPC was located on human chromosome 12. In situ hybridization of the same genomic PRP probe pinpointed the SPC to chromosome band 12p13.2 in humans [82]. Somatic cell hybrid studies in mice originally suggested that PRP genes were located on chromosome 8 [83] but more recently work with recombinant inbred strains has demonstrated linkage with markers on the distal end of chromosome 6 [84]. The authors note that the chromosome 6 location is more consistent with conservation of linkage groups between mice and humans.

The number of genes coding for the plethora of PRP polymorphisms has long been an issue. The similarities in biochemical properties, electrophoretic phenotypes and immunological relationships of the human PRP [85] led Friedman and Karn [86] to propose that these proteins could result from the cleavage of a core protein by the cascade action of two proteolytic enzymes. Such a posttranslational derivation of some or all of the human PRP would reduce the number of genes required to encode them. They subsequently demonstrated that the smaller members, Pr 3 and Pr 4, in the acidic PRP phenotypes could be derived from the larger members, Pr 1 and Pr 2, by posttranslational modification [87, 88]. They further demonstrated that this modification was probably due to the presence of a kallikrein in parotid saliva. Comparison of peptides stained specifically for arginine with the molar value for arginine in the protein suggested that the enzyme's action was to cleave the larger version at $arg^{106}$. That proposal was later confirmed by Wong et al. [89]. Other peptide studies done in Karn's laboratory with limited proteolysis suggested that a posttranslational cleavage relationship also exists between Ps 1 and 2 [90], extending data on the posttranslational derivation of the smaller peptides to the basic PRP.

Mamula et al. [91] observed that PRP mRNAs were larger than predicted from the sizes of the proteins, consistent with the core protein hypothesis of Friedman and Karn [86]. They observed PRP mRNAs in the range of 1.1–1.8 kb which are large enough to encode proteins between 20,000 and 35,000 daltons. Subsequently, they demonstrated six major PRP products of cell-free translation of those mRNAs using $^{3}H$-proline [92]. The corrected molecular weights were between 10,000 and 35,000 daltons. Extensive molecular analyses of PRP cDNAs and genes (see chapter by Ferguson, pp. 95–116) have confirmed the posttranslational hypothesis originally advanced by Friedman and Karn and extended it to include exon shuffling as a means of accounting for some of the complexity of this family of salivary proteins. Maeda [93] reinterpreted Azen's data to suggest that Pa, Db and PIF represented alleles of a single locus, a hypothesis supported by later molecular studies. She proposed that six genes encode the human parotid PRP.

At this time, cDNAs have been produced and sequenced from rat [94], mouse [95] and humans [96]. All contain repeated sequences as expected from the amino acid sequences [for reviews see 7, 8] of the proteins they represent. Various studies suggest that six genes located on the short arm of chromosome 12 encode the human PRP and that two genes [97] located on chromosome 6 encode the mouse PRP. The expression of these genes is quite different in these two organisms. Human PRP are expressed constitutively while rodent PRP are inducible. One expects that the difference in expression may be due to different roles for these proteins in human and rodent salivas. The results of function studies suggest that the acidic human PRP have an important role in maintaining the supersaturated level of calcium in saliva and preventing formation of hydroxyapatite crystals on tooth surfaces [7, 8]. It has been proposed that they play an important part in formation of dental pellicle by binding to hydroxyapatite [7, 8]. By comparison, this involvement may be unimportant or nonexistent for rodent PRP which are not constitutively produced.

Based on observations of the affinity of proline-rich sequences for polyphenols, Carlson and co-workers [15, 20–22] have proposed that these proteins are responsible for binding and neutralization of dietary tannins.

They have compared mammalian PRP to a similar PRP contained in a water-soluble mucilage produced by fungi that grow on cereal crops [22]. This fungal protein has a very high affinity for tannins and, when it is absent, these fungal spores do not germinate in the presence of phenolic compounds. Recently, others have proposed that human PRP may have a similar role, citing high consumption of tannin-rich foods by populations at high risk for esophageal carcinoma [98]. They propose that human PRP protect the esophagus from carcinogenic effects by binding tannins. Since a function(s) has not been assigned to the basic human PRP, one could speculate that they may be especially good candidates for this. The authors note extensive vegetable sources of tannins in the diets of populations in Central Asia, Northern China and South Africa and compare their incidence of esophageal carcinoma with those of populations from North America and Western Europe where increased risk may be due to different factors such as extensive consumption of alcohol and tobacco products.

### *Statherin Genetics*

The proline-rich salivary protein statherin is less abundant than the members of the PRP family described above and differs from them in that it is also tyrosine-rich [10, 99, 100]. Using an antibody specific for human statherin, Oppenheim et al. [100] immunoprecipitated a statherin precursor from cell-free translations of human submandibular gland poly(A+)RNA and showed that it had a molecular weight of 7,800, about 2,000 higher than native statherin. Since that study, the statherin cDNA has been characterized [101] and shown to determine a peptide consistent with the sequence originally determined by Schlesinger and Hay [102]. Thus the posttranslational proteolytic cleavages that produce smaller versions of PRP do not appear to affect the final product of the statherin gene as is the case with PRP gene products (see above). Statherin appears to be encoded by a single gene on the human chromosome segment 4q11-13 [101]. Statherin binds calcium and, in conjunction with acidic PRP, helps maintain the supersaturation of calcium phosphate salts in saliva. This in turn prevents precipitation of those salts onto tooth surfaces and in the salivary glands [103]. Statherin binds to hydroxyapatite surfaces [7, 10] and, in conjunction with PRP, can promote the attachment of *Actinomyces* to apatitic surfaces [104]. RNA hybridizing with a statherin DNA probe has been found in human parotid gland, submandibular gland and bronchus [105]. The same study showed hybridization of the probe to RNA in the parotid, submandibular and lacrimal glands of the rhesus monkey.

### *Histatin Genetics*

Histidine-rich proteins (HRP) or histatins, are a series of histidine-rich phosphoproteins [11] that appear to be involved in the nonimmune defense system in the oral cavity by providing bactericidal [17] and fungicidal activities [11, 106, 107] to saliva. Histatin 1, also known as neutral HRP, binds to hydroxyapatite and participates in the inhibition of crystal growth [11] along with the PRP described above. The genes for histatins are probably the same as those that encode Pb (parotid basic protein) and Pbb (post-Pb) [108]. A study of 12 cDNA clones obtained from a human parotid library with mixed oligonucleotide probes representing a common coding region, suggests that histatin mRNAs are 95% homologous and that they are encoded by at least two genes, *His1* and *His2* [109]. The investigators conclude that *His1* determines histatin 1 and *His2* determines histatin 3. They hypothesize that histatins 4–6 are derived from histatin 3 by proteolytic cleavages and that histatin 2 represents a nonphosphorylated form of histatin 1. Experiments using a fragment of the histatin 3 cDNA have shown that histatin expression is restricted to the parotid and submandibular glands [105].

### *Cystatin Genetics*

Cystatins are a family of cysteine-rich salivary phosphoproteins [12] that have been characterized as cysteine protease inhibitors [19]. In general, cysteine proteinase inhibitors inhibit ficin, papain, and the cathepsins with varying specificities [110, 111]. Like other salivary phosphoproteins, the cystatins bind to hydroxyapatite and inhibit crystal growth [12]. The genes encoding the human salivary cystatins are part of a multigene family composed of as many as seven members [112, 113], some of which may code for nonsalivary proteins. Three of these genes have been characterized and two of them identified with inhibition activities for different cathepsins. These two genes appear to be evolutionarily related [114]. The third appears to be a pseudogene [113]. Cystatin mRNAs have been detected in human bronchus, submandibular glands and parotid glands with substantially more expression in the submandibular glands than in the parotid glands [105].

### *Vitamin $B_{12}$-Binding Protein Genetics*

Polymorphism of a vitamin $B_{12}$-binding protein that appears in parotid saliva, tears, milk and leukocytes has been described [115]. Significant differences in the distributions of two codominant alleles were seen in Whites, Blacks and Chinese. The gene showed no linkage to the PRP genes of the SPC and no chromosomal assignment is yet available.

### *Parotid Secretory Protein Genetics*

In addition to salivary amylase (see previous discussion), mouse parotid glands have been shown to synthesize large quantities of a protein designated parotid secretory protein (PSP) [13]. Leucine accounts for 49 of 215 residues of the putative protein predicted from the base sequence of the cloned cDNA [13]. Two alleles have been described which differ in the methionine content of their products: one from the YBR strain which had one to two residues of methionine and one from the C3H strain which had no methionine [116]. PSP mRNA has been detected in the parotids of other rodents and in the parotid and submandibular glands of man [117].

The PSP gene has been mapped to mouse chromosome 2 [116] where it exists in a single copy [117]. Control of expression of PSP and salivary amylase, which also exists in only a single copy in most mice (see earlier discussion), appears to be tightly coordinated in fully differentiated parotid glands [116, 117] even though they are on different chromosomes. Developmental studies have shown that this is not the case during ontogeny when increases in the two proteins are disproportionate [117]. A model has been proposed in which both genes are subject to transcriptional regulation by factors affecting their promotors in the salivary gland, and while their rates of induction are not comparable, the ratios of their final expression levels are fixed [117].

### *Androgen-Binding Protein Genetics*

Some years ago it was observed that mouse saliva contains a type of protein capable of binding testosterone. The protein was designated androgen-binding protein (ABP) [16] and, more recently, ABP has been shown to be a series of proteins that are genetically and structurally related [118]. A survey of a variety of wild rodents, including representatives in the Muridae and Cricetidae, have shown salivary ABP to be widespread among this order [Karn, unpubl. data].

The primary source of mouse ABP is the submandibular gland although they have been found in very small quantities in the parotid and sublingual glands [119]. They are dimers composed of disulfide-bridged subunits of two different sizes, alpha (ca. 12,000 daltons) and beta (ca. 8,800 daltons) [16]. The amino acid compositions of ABP have been determined and are not particularly unusual, although they are relatively rich in dicarboxylic acids and in leucine, alanine and lysine [119]. No evidence for glycosylation was found in those studies.

Interesting structural correlations can be made between mouse ABP and the major secretory protein of rat prostate, called prostatic steroid-binding protein (PBP) [120]. PBP is composed of two disulfide-bridged dimers whose subunits have molecular weights similar to those of the subunits of ABP [16, 119]. Investigations of mouse tissue extracts and fluids with an antibody specific for ABP have shown that the only other tissue containing cross-reactive material (CRM) is the prostate [119]. While it is tempting to speculate that the prostate CRM is the mouse version of PBP, that does not seem likely. The quantity produced in the mouse prostate and lack of cross-reactivity of the anti-ABP antibody with rat prostate extracts [Karn, unpubl. data] are at variance with expectations for a relationship between the two rodent proteins.

Genetic studies suggest that, in addition to alpha and beta subunits, there is a third subunit with a molecular weight comparable to that of beta, which has been termed the gamma subunit [118]. Gamma appears to be an alternative to beta in that it is always found with alpha in a dimer. It appears, however, that much less gamma is made than either alpha or beta. Two variants of the alpha ($A^a$ and $A^b$) and gamma ($G^a$ and $G^b$) subunits were identified in inbred strains of mice; no variant of beta has yet been seen [118]. These findings suggest that there might be as many as three structural genes for ABP, one for each subunit. That hypothesis was supported by the finding of three products of cell-free translation of submandibular RNA which could be precipitated with the anti-ABP antibody [119]. That study also showed that at least the alpha *(Abpa)* and gamma *(Abpg)* genes are closely linked on the proximal end of mouse chromosome 7. The location of the beta gene *(Abpb)* is not currently known.

An intriguing aspect of the mouse salivary ABP phenotypes is that their electrophoretic mobilities can be changed by the presence of particular alleles of modifier genes. The first of these to be reported was *sex-limited saliva pattern (SSP)* with dominant allele *Ssp*$^{S}$ whose action reduces the anodal mobility of ABP in male salivas [121]. The expression of *Ssp*$^{S}$ is dependent on the increased levels of androgen that occur in males following puberty [121, 122] and upon the presence of an intact androgen receptor [16]. As a result, the apparent molecular weight of the alpha subunit is increased while the charge apparently changes little [16]. The molecular weight increase appears to be due to the addition of covalently bonded molecules to the alpha subunit. This does not happen when ABP binds androgen and apparently is not the result of glycosylation [16, 119]. Recently a modifier gene with similar but somewhat different action has been identified in inbred *Mus musculus castaneus* [Karn, unpubl. data]. Like *Ssp*$^{S}$, the dominant allele of this new gene (which has yet to be named) causes an increase in the apparent size of the alpha subunit. The result, however, is to increase the anodic mobility of the alpha subunit, suggesting that the charge of the subunit has been significantly increased.

At this juncture the complex ABP system comprises a multitude of mysteries. Why does mouse saliva contain a protein capable of binding male sex hormones? If the biological function is related to its steroid-binding capability, two possibilities suggest themselves: general binding/neutralization of dietary substances (toxins, nutrients, etc.) with structures related to steroids or, more specifically, binding sex steroids for oral and/or olfactory recognition purposes [123]. Why do genes exist which can modify the molecular weight and, in some cases, the charge of the alpha subunit? Why is the synthesis of ABP apparently not under hormonal control but the expressions of the modification genes are? Why are the modifier genes polymorphic in mouse populations? The answers to these questions must await further analyses of the binding properties of the proteins, but one could speculate that the modifications may alter the binding properties of these proteins in subtle ways that are important to their function in one or more environments.

### *Submandibular Renin Genetics*

Renin is an aspartyl protease that catalyzes the initial, rate-limiting step in the conversion of angiotensinogen to the vasoactive hormone angiotensin II. The enzyme is usually produced in highest quantity in the kidney from where it is contributed to the bloodstream and plays a key role in the regulation of salt water balance and blood pressure regulation [see 124 for a review]. Low concentrations of renin can be found in other tissues but most do not produce it in quantity. An exception are the submandibular glands of some strains of mice where renin is sometimes produced in quantities rivaling that of the kidney. Renin genes are found on chromosome 1 in mice and whether they produce low or high quantities of submandibular renin depends on whether they have one or two gene copies. Rats have only one renin gene which is expressed primarily in the kidney. Renin transcripts are not found at all in rat submandibular glands [125]. Of the two mouse renin genes, the $Ren^1$ gene is found in all strains and is expressed primarily in the kidney. The $Ren^2$ gene appears in the strains capable of producing high levels of submandibular renin and is responsible for most of that production [126].

### *Regulation of Salivary Protein Gene Expression*

In this chapter, I have attempted to compare and contrast the variety of unusual proteins that appear in mammalian salivas as the result of the expressions of salivary gland genes. Many of these are tissue-specific as documented by Sabatini et al. [105]. Regulation of gene expression in eukaryotes appears to involve both positive and negative *cis*-acting regulatory DNA sequences [127]. Such sequences have not yet been absolutely identified for salivary protein genes but Gumucio et al. [55] have compared the upstream sequences of several mouse and human salivary protein genes to identify consensus sequences that might be candidates for such roles. Specifically, they found several shared sequences between the promoter of the human salivary amylase gene they sequenced and the promoter of a human PRP gene sequenced by others [128]. In this context, it is interesting to note that Wright and Carlson [129] observed coordinate-positive stimulation of both alpha-amylase and PRP by dibutyryl cyclic AMP and forskolin in some of the hybrid cell lines that they produced by fusing rat parotid gland cells with a rat hepatoma cell line.

Perhaps one of the more puzzling aspects of tissue-specific expression involving a salivary gland protein gene is the expression of amylase in mammalian livers. Some time ago it was determined that, contrary to speculation at the time, there was no convincing evidence for a unique amylase produced in the human liver and which was contributed to the serum [56]. While that has proved to be true, it has since been shown that the mouse salivary amylase gene is expressed in low levels in liver [31] and that the protein can be contributed to the serum, at least in some cases [44]. This is due to a liver-specific promoter lying upstream of the mouse structural gene, between it and the salivary promoter [31]. The coding sequence, however, produces a salivary-type amylase. Recently it has been shown that mouse liver also expresses the pancreatic amylase gene [44]. Apparently mouse liver is capable of contributing both salivary and pancreatic amylases to the serum.

In humans, on the other hand, amylase expression occurs in the liver due to the low-level expression of one of the two pancreatic amylase genes [32]. Again, expression in the liver appears to be due to a promoter distinct from that used in the pancreas for expression of that gene. Why amylase is produced in the liver in these mammals remains a mystery. Particularly intriguing is the finding that mouse liver expression is both salivary and pancreatic amylase while human liver expression is pancreatic amylase only. Are these low-level gene expressions in the liver vestiges of an evolutionarily earlier function? If they represent a current function, does it matter what type of amylase is expressed in this tissue? These questions may be important for a better understanding of subtle and complex digestive functions.

## *Conclusions*

Research on salivary proteins has come a long way from a time when saliva was viewed as a relatively simple mixture consisting primarily of soluble salts and amylolytic enzyme activity. Much of this progress is due to the ease with which relatively large amounts of salivary proteins can be obtained by noninvasive methods. Today we view mammalian salivas as complex mixtures containing some very unusual proteins. These proteins confer properties on saliva that are important for its participation in the various functions of the oral cavity. Some of these properties are apparent and some only speculative at this time. Examples of salivary proteins whose functions are obscure include PSP and ABP produced by mouse parotid and submandibular glands, respectively. More importantly, salivary proteins probably have other functions that have not yet been identified and some of these may involve more than one type of protein operating in a synergistic fashion. An example is their role in the complex deposit, plaque, which involves not only salivary proteins but particular microorganisms. Perhaps one of the most intriguing pictures developing from the many and diverse studies of mammalian salivary proteins is that forming from the comparison of proteins in the salivas in different mammals. Why do humans produce PRP constitutively while the equivalent proteins in rodents appear only under certain conditions? Why do mice express renin genes in their salivary glands and rats do not? Why do mouse salivas contain proteins such as PSP and ABP that are not found in human saliva?

Clearly, studies of salivary proteins such as amylase and PRP have contributed much to our understanding of multiple gene families and the expression of their products. The origins of a wide variety of proteins in saliva involve complex interactions, including exon shuffling in the case of the PRP and posttranslational modifications such as proteolytic cleavage, phosphorylation, glycosylation and deamidation in the PRP, amylase and others. One of the most intriguing and poorly understood mechanisms that produces variation in salivary proteins is the interaction of androgen-regulated modifier genes with the structural gene determining the mouse salivary ABP alpha subunit. The purpose of these polymorphic genes, that increase the apparent molecular weight and, in some cases, the charge of this protein, remains a mystery.

Studies of tissue localization of the messenger RNAs for these proteins and the distribution of the proteins themselves have developed a picture of the relative specificity of expression of their genes. Some proteins originally described in saliva obviously have wider distribution and clarification of this distribution promises to enhance our understanding of the overall function(s) of the proteins themselves. Studies of coordinate control of the expression of salivary proteins also help in this regard and add measurably to our general understanding of the regulation of mammalian gene expression.

## *References*


1 Karn RC, Malacinski GM: The comparative biochemistry, physiology, and genetics of animal α-amylases. Adv Comp Physiol Biochem 1978;7:1–103.
2 Merritt AD, Karn RC: The human α-amylases. Adv Hum Genet 1977;8:135–234.
3 Iwamoto Y, Nakamura R, Tsunemitsu A, et al: The heterogeneity of human salivary peroxidase. Arch Oral Biol 1968;13:1015–1021.
4 Eichel HJ, Conger N, Chernick WS: Acid and alkaline ribonucleases of human parotid, submaxillary, and whole saliva. Arch Biochem Biophys 1964;107:197–208.
5 Modeer T: Human saliva kallikrein. Acta Odontol Scand 1977;35:23–32.
6 Azen EA: Genetic protein polymorphisms in human saliva: An interpretive review. Biochem Genet 1978;16:79–99.
7 Bennick A: Salivary proline-rich proteins. Mol Cell Biochem 1982;45:83–99.
8 Bennick A: Structural and genetic aspects of proline-rich proteins. J Dent Res 1987; 66:457–461.

9 Azen EA, Maeda N: Molecular genetics of human salivary proteins and their polymorphisms. Adv Hum Genet 1988;17:141–199.
10 Hay DI: The isolation from human parotid saliva of a tyrosine-rich acidic peptide which exhibits high affinity for hydroxyapatite surfaces. Arch Oral Biol 1973;18:1531–1541.
11 Oppenheim FG, Yang Y-C, Diamond RD, et al: The primary structure and functional characterization of the neutral histidine-rich polypeptide from human parotid secretion. J Biol Chem 1986;261:1177–1185.
12 Shomers JP, Tabak LA, Levine MJ, et al: Properties of cysteine-containing phosphoproteins from human submandibular-sublingual saliva. J Dent Res 1982;61:397–399.
13 Madsen HO, Hjorth JP: Molecular cloning of mouse PSP mRNA. Nucleic Acids Res 1985;13:1–13.
14 Grasbeck R, Visuri K, Stenman U: The vitamin $B_{12}$-binding protein in human saliva: Isolation, characterization and comparison with unpurified protein. Biochim Biophys Acta 1972;263:721–733.
15 Mehansho H, Hagerman A, Clements S, et al: Modulation of proline-rich protein biosynthesis in rat parotid glands by sorghums with high tannin levels. Proc Natl Acad Sci USA 1983;80:3948–3952.
16 Dlouhy SR, Karn RC: The tissue source and cellular control of the apparent size of androgen binding protein (Abp), a mouse salivary protein whose electrophoretic mobility is under the control of *sex-limited saliva pattern (Ssp).* Biochem Genet 1983;21:1057–1070.
17 MacKay BJ, Denepitiya L, Iacono VJ, et al: Growth-inhibitory and bactericidal effects of human parotid salivary histine-rich polypeptides on *Streptococcus mutans.* Infect Immunol 1984;44:695–701.
18 Xu T, Levitz SM, Oppenheim FG: Fungistatic and fungicidal activity of parotid saliva proteins. J Dent Res 1988;67:236.
19 Isemura S, et al: Cystatin S and the related cysteine proteinase inhibitors in human saliva; in Turk V (ed): Cysteine Proteinases and Their Inhibitors. Berlin, de Gruyter, 1986, pp 497–505.
20 Carlson DM, Ann DK, Mehansho H: Proline-rich proteins: Expression of salivary multigene families; in Leive (ed): Microbiology. American Society for Microbiology, Washington DC, 1986, pp 303–306.
21 Mehansho H, Butler LG, Carlson DM: Dietary tannins and salivary proline-rich proteins: Interactions, induction and defense mechanisms. Annu Rev Nutr 1987;7: 423–440.
22 Carlson DM: Proline-rich proteins and glycoproteins: Expressions of salivary gland multigene families. Biochimie 1988;70:1689–1695.
23 Azen EA, Lush IE, Taylor B: Close linkage of mouse genes for salivary proline-rich proteins and taste. Trends Genet 1986;2:199–200.
24 Lush IE, Holland G: The genetics of tasting in mice. V. Glycine and cyclohexamide. Genet Res (Cambridge) 1988;52:207–212.
25 O'Hanlon K, Weissbecker K, Cortessis V, et al: Genes for salivary proline-rich proteins and taste for phenylthiourea are not closely linked in humans. Cytogenet Cell Genet 1988;49:315–317.
26 Meisler MH, Gumucio DL: Salivary amylase: evolution and tissue-specific expression; in Desnuelle P, Sjöström H, Noren O (eds): Molecular and Cellular Basis of Digestion. Amsterdam, Elsevier/North-Holland, 1986, pp 457–466.
27 Meisler MH, Gumucio DL: Pancreatic amylase: molecular genetics and evolution; in

Desnuelle P, Sjostrom H, Noren O (eds): Molecular and Cellular Basis of Digestion. Amsterdam, Elsevier/North-Holland, 1986, pp 249–263.
28 Schneyer LH: Amylase content of separate salivary gland secretions of man. J Appl Physiol 1956;9:453–455.
29 Skude G, Mardh PA, Westrom L: Amylases of the genital tract. I. Isoamylases of genital tract tissue homogenates and peritoneal fluid. Am J Obstet Gynecol 1976; 126:652–656.
30 Tosi M, Young RA, Hagenbuchle O, et al: Multiple polyadenylation sites in a mouse α-amylase. Nucleic Acids Res 1981;9:2313–2323.
31 Hagenbuchle O, Tosi M, Schibler U, et al: Mouse liver and salivary gland α-amylase mRNAs differ only in 5′ nontranslated sequences. Nature 1981;289:643–646.
32 Samuelson LC, Wiebauer K, Gumucio DL, et al: Expression of the human amylase genes: Recent origin of a salivary amylase promoter from an actin pseudogene. Nucleic Acids Res 1988;16:8261–8276.
33 Karn RC, Petersen TE, Hjorth JP, et al: Characterization of the amino termini of mouse salivary and pancreatic amylases. FEBS Lett 1981;126:292–296.
34 Carne T, Scheele G: Amino acid sequences of transport peptides associated with canine exocrine pancreatic proteins. J Biol Chem 1982;257:4133–4140.
35 Fischer EH, Stein EA: Alpha-amylases; in Boyer PD (ed): The Enzymes, ed 2. New York, Academic Press, 1960, vol 4, pp 313–343.
36 Karn RC, Rosenblum BB, Ward JC, et al: Genetic and post-translational mechanisms determining human amylase isozyme heterogeneity; in Markert CL (ed): Isozymes, vol IV: Genetics and Evolution. New York, Plenum Press, 1977, pp 745–761.
37 Sick K, Nielsen JT: Genetics of amylase isozymes in the mouse. Hereditas 1964; 51:291–296.
38 Nielsen JT: Variation in the number of genes coding for salivary amylase in the bank vole, *Clethrionomys glareola*. Genetics 1977;85:155–169.
39 Hagenbuchle O, Bovey R, Young RA: Tissue-specific expression of mouse α-amylase genes: Nucleotide sequence of isoenzyme mRNAs from pancreas and salivary gland. Cell 1980;21:179–187.
40 Wise RJ, Karn RC, Larsen SH, et al: A complementary DNA sequence that predicts a human pancreatic amylase primary structure consistent with the electrophoretic mobility of the common isozyme, $Amy_2$ A. Mol Biol Med 1984;2:307–322.
41 Dawson WD, Huang LL: Comparative genetics of hamster amylases. Biochem Genet 1981;19:623–633.
42 Nielsen JT: Genetic studies of the amylase isoenzyme of the bank vole, *Clethrionomys glareola*. Hereditas 1969;61:400–412.
43 Schibler U, Pittet A-C, Young RA, et al: The mouse α-amylase multigene family: sequence organization of members expressed in the pancreas, salivary gland and liver. J Mol Biol 1982;155:247–266.
44 Jones JM, Keller SA, Samuelson LC, et al: A salivary amylase transgene is efficiently expressed in liver but not in parotid gland of transgenic mice. Nucleic Acids Res 1989;17:6613–6623.
45 Nielsen JT: Variations in amylase haplotypes among congenic lines of the house mouse. Genetics 1982;102:571–582.
46 Strahler JR, Meisler M: Two distinct pancreatic amylase genes are active in YBR mice. Genetics 1982;101:91–102.

47 Mikkelsen BM, Clark ME, Christiansen G, et al: The structure of two distinct pancreatic amylase genes in mouse strain YBR. Biochem Genet 1985;23:511–524.
48 Gumucio DL, Wiebauer K, Dranginis A, et al: Evolution of the amylase multigene family. J Biol Chem 1985;260:13483–13489.
49 Tosi M, Bovey R, Astolfi S, et al: Multiple nonallelic genes encoding pancreatic alpha-amylase of mouse are expressed in a strain-specific fashion. EMBO J 1984;3: 2809–2816.
50 Pronk JC, Frants RR, Jansen W, et al: Evidence for duplication of the human salivary amylase gene. Hum Genet 1982;60:32–35.
51 Nakamura Y, Ogawa M, Nishide T, et al: Sequences of cDNAs for human salivary and pancreatic α-amylases. Gene 1984;28:263–270.
52 Nishide T, Emi M, Nakamura Y, et al: Corrected sequences of cDNAs for human salivary and pancreatic α-amylases. Gene 1986;50:371–372.
53 Nishide T, Nakamura Y, Emi M, et al: Primary structure of human salivary α-amylase gene. Gene 1986;41:299–304.
54 Handy DE, Larsen SH, Karn RC, et al: Identification of a human salivary amylase gene. Mol Biol Med 1987;4:145–155.
55 Gumucio DL, Wiebauer K, Caldwell RM, et al: Concerted evolution of human amylase genes. Mol Cell Biol 1988;8:1197–1205.
56 Karn RC, Wise RJ, Merritt AD: The tissue origins of serum and urinary α-amylase. Arch Biochem Biophys 1976;175:144–152.
57 Groot PC, Bleeker MJ, Pronk JC, et al: The human alpha amylase multigene family consists of haplotypes with variable numbers of genes. Genomics 1989;5:29–42.
58 Zabel B, Naylor S, Sakaguchi A, et al: High resolution chromosomal localization of human genes for amylase, proopiomelanocortin, somatostatin and a DNA fragment (D3S1) by in situ hybridization. Proc Natl Acad Sci USA 1983;80:6932–6936.
59 Eicher EM, Lane PW: Assignment of LG XVI to chromosome 3 in the mouse. J Hered 1980;71:315–318.
60 Mandel ID, Thompson RH, Ellison SA: Studies on the mucoproteins of human parotid saliva. Arch Oral Biol 1965;10:499–507.
61 Warner T, Azen EA: Proline-rich proteins are present in serous cells of submucosal glands in the respiratory tract. Am Rev Resp Dis 1984;130:115–118.
62 Ito S, Isemura S, Saitoh E, et al: Immunohistochemical demonstration of salivary proline-rich peptide P-C-like immunoreactivity in human pancreatic β-cells. Acta Endocrinol 1983;104:544–551.
63 Muenzer J, Bildstein C, Gleason M, et al: Purification of proline-rich proteins from parotid glands of isoproterenol-treated rats. J Biol Chem 1979;254:5623–5628.
64 Muenzer J, Bildstein C, Gleason M, et al: Properties of proline-rich proteins from parotid glands of isoproterenol-treated rats. J Biol Chem 1979;254:5629–5634.
65 Mehansho H, Clements S, Sheares BT, et al: Induction of proline-rich glycoprotein synthesis in mouse salivary glands by isoproterenol and by tannins. J Biol Chem 1985;260:4418–4423.
66 Carlson DM, Mehansho H, Ann DK: Induction of proline-rich protein biosynthesis in hamster salivary glands by isoproterenol treatment. Fed Proc 1985;44:1097.
67 Azen EA, Oppenheim F: Genetic polymorphism of proline-rich human salivary proteins. Science 1973;180:1067–1073.
68 Friedman RD, Merritt AD, Rivas ML: Genetic studies of human acidic salivary protein (Pa). Am J Hum Genet 1974;27:292–303.

69 Azen EA, Denniston CL: Genetic polymorphism of human salivary proline-rich proteins: Further genetic analysis. Biochem Genet 1974;12:109–120.
70 Azen EA, Hurley CK, Denniston CL: Genetic polymorphism of the major parotid salivary glycoprotein (Gl) with linkage to the genes for Pr, Db, and Pa. Biochem Genet 1979;17:257–279.
71 Ikemoto S, Minaguchi K, Suzuki K, et al: New genetic marker in human parotid saliva (Pm). Science 1977;197:378–379.
72 Azen EA, Denniston CL: Polymorphism of Ps (parotid size variant) and detection of a protein (PmS) related to the Pm (parotid middle band) system with genetic linkage of *Ps* and *Pm* to *Gl*, *Db*, and *Pr* genetic determinants. Biochem Genet 1980;18:483–501.
73 Azen EA, Denniston CL: Genetic polymorphism of PIF (parotid isoelectric focusing variant) proteins with linkage to the PPP (parotid proline-rich protein) gene complex. Biochem Genet 1981;19:475–485.
74 Azen EA, Yu P-L: Genetic polymorphism of CON-1 and CON-2 salivary proteins detected by immunologic and concanavalin A reaction on nitrocellulose with linkage of CON-1 and CON-2 genes to the SPC (salivary protein complex). Biochem Genet 1984;22:1–19.
75 Azen EA, Yu P-L: Genetic polymorphism of Pe and Po salivary proteins with probable linkage of their genes to the SPC (salivary protein complex). Biochem Genet 1984;22:1065–1080.
76 Karn RC, Goodman PA, Yu P-L: Description of a genetic polymorphism of a human proline-rich salivary protein, Pc, and its relationship to other proteins in the salivary protein complex, SPC. Biochem Genet 1985;23:37–51.
77 Minaguchi K, Suzuki K: New salivary protein polymorphisms detected by SDS polyacrylamide gel electrophoresis. J Dent Res 1986;63:326.
78 Goodman PA, Yu P-L, Azen EA, et al: The human salivary protein complex (SPC): A large block of related genes. Am J Hum Genet 1985;37:785–797.
79 Goodman PA, Yu P-L, Azen EA, et al: Probable location of a large block of salivary protein genes on human chromosome six. Am J Hum Genet 1982;34:182A.
80 Goodman PA, Lalley PA, Azen EA: Salivary proline-rich protein genes on chromosome 12. Am J Hum Genet 1984;36:202s.
81 Azen EA, Goodman PA, Lalley PA: Human salivary proline-rich protein genes on chromosome 12. Am J Hum Genet 1985;37:418–424.
82 Mamula PW, Heerema NA, Palmer CG, et al: Localization of the human salivary protein complex (SPC) to chromosome band 12p13.2. Cytogenet Cell Genet 1985; 39:279–284.
83 Azen EA, Carlson DM, Clements S, et al: Salivary proline-rich protein genes on chromosome 8 of mouse. Science 1984;226:967–969.
84 Azen EA, Davisson MT, Cherry M, et al: Prp (proline-rich protein) genes linked to markers ES-12 (esterase-12), Ea-10 (erythrocyte alloantigen), and loci on distal mouse chromosome 6. Genomics 1989;5:415–422.
85 Friedman RD, Karn RC: Immunological relationships and a genetic interpretation of major and minor acidic proteins in human parotid saliva. Biochem Genet 1977; 15:549–562.
86 Friedman RD, Karn RC: An hypothetic mechanism concerning the origin of proline-rich human salivary proteins. Am J Hum Genet 1977;29:44A.
87 Karn RC, Friedman RD, Laughner W, et al: Purification and partial characterization of human salivary proline-rich proteins. Fed Proc 1977;36:739.

88 Karn RC, Friedman RD, Merritt AD: Human salivary proline-rich (Pr) proteins: A posttranslational derivation of the phenotypes. Biochem Genet 1979;17:1061–1077.
89 Wong RSC, Madapallimattam G, Bennick A: The role of glandular kallikrein in the formation of a salivary proline-rich protein A by cleavage of a single bond in salivary protein C. Biochem J 1983;211:35–44.
90 Goodman PA, Karn RC: Human parotid size polymorphism (Ps): Characterization of two allelic products, Ps 1 and 2, by limited proteolysis. Biochem Genet 1983;21: 405–416.
91 Mamula PW, Wise RJ, Karn RC: Isolation and characterization of human parotid RNA. Fed Proc 1984;43:1522.
92 Mamula PW, Morley DJ, Larsen SH, et al: Expression of human salivary protein genes. Biochem Genet 1988;26:165–175.
93 Maeda N: Inheritance of the human salivary proline-rich proteins: A reinterpretation in terms of six loci forming two subfamilies. Biochem Genet 1985;23:455–464.
94 Ziemer M, Swain W, Rutter W, et al: Nucleotide sequence analysis of a proline-rich protein cDNA and peptide homologies of rat and human proline-rich proteins. J Biol Chem 1984;259:10475–10480.
95 Clements S, Mehansho H, Carlson DM: Novel multigene families encoding highly repetitive peptide sequences. J Biol Chem 1985;260:13471–13477.
96 Maeda N, Kim H-S, Azen EA, et al: Differential RNA splicing and posttranslational cleavages in the human salivary proline-rich protein gene system. J Biol Chem 1985; 260:11123–11130.
97 Ann DK, Smith MK, Carlson DM: Molecular evolution of the mouse proline-rich protein multigene family: Insertion of a long interspersed repeated DNA element. J Biol Chem 1988;263:10887–10893.
98 Warner TF, Azen EA: Tannins, salivary proline-rich proteins and cancer. Med Hypotheses 1988;26:99–102.
99 Oppenheim FG, Offner GD, Troxler RF: Phosphoproteins in the parotid saliva from the subhuman primate *Macaca fasicularis*. J Biol Chem 1982;257:9721–9782.
100 Oppenheim FG, Hay DI, Smith DJ, Offner GD, Troxler RF: Molecular basis of salivary proline-rich protein and peptide synthesis: Cell-free translations and processing of human and macaque statherin mRNAs and partial amino acid sequence of their signal peptides. J Dent Res 1987;66:462–466.
101 Sabatini LM, Carlock LR, Johnson GW, et al: cDNA cloning and chromosomal location of a gene (4q11-13) for statherin, a regulator of calcium in saliva. Am J Hum Genet 1987;41:1048–1060.
102 Schlesinger DH, Hay DI: The complete covalent structure of statherin, a tyrosine-rich acidic peptide which inhibits calcium phosphate precipitation from human parotid saliva. J Biol Chem 1977;252:1689–1695.
103 Hay DI, Smith DJ, Schluckebier SK, et al: Relationship between concentration of human salivary statherin and inhibition of calcium phosphate precipitation in stimulated human parotid saliva. J Dent Res 1984;63:857–863.
104 Gibbons RJ, Hay DI: Human salivary acidic proline-rich proteins and statherin promote the attachment of *Actinomyces viscosus* LY7 to apatitic surfaces. Infect Immun 1988;56:439–445.
105 Sabatini LM, Warner TF, Saitoh E, et al: Tissue distribution of RNAs for cystatins, histatins, statherin, and proline-rich proteins in humans and macaques. J Dent Res 1969;68:1138–1145.

106 Pollock JJ, Denepitiya L, MacKay BJ, et al: Fungistatic and fungicidal activity of human parotid salivary histidine-rich polypeptides on *Candida albicans*. Infect Immunol 1984;44:702–707.

107 Oppenheim FG, Xu T, McMillian FM, et al: Histatins, a novel family of histidine-rich proteins in human parotid secretion. Isolation, characterization, primary structure and fungistatic effects on *Candida albicans*. J Biol Chem 1988;263:7472–7477.

108 Peters EH, Azen EA: Isolation and partial characterization of human parotid basic proteins. Biochem Genet 1977;15:925–946.

109 Sabatini LM, Azen EA: Histatins, a family of salivary histidine-rich proteins, are encoded by at least two loci *(His1* and *His2)*. Biochem Biophys Res Commun 1989; 160:495–502.

110 Green GDJ, Kembhavi AA, Davies ME, et al: Cystatin-like cysteine proteinase inhibitors from human liver. Biochem J 1984;218:939–946.

111 Minakata K, Asano M: New protein inhibitors of cysteine proteinases in human saliva and salivary glands. Hoppe Seylers Z Physiol Chem 1984;365:399–403.

112 Al-Hashimi I, Dickinson DP, Levine MJ: Molecular cloning of human salivary cystatin SA-1. J Biol Chem 1988;263:9381–9387.

113 Saitoh E, Kim H-S, Smithies O, et al: Human cysteine-proteinase inhibitors: Nucleotide sequence analysis of three members of the cystatin gene family. Gene 1987;61: 329–338.

114 Saitoh E, Isemura S, Sanada K, et al: Cystatin superfamily: Evidence that family II cystatin genes are evolutionarily related to family III cystatin genes. Biol Chem Hoppe Seyler 1988;369 (suppl):191–197.

115 Azen EA, Denniston C: Genetic polymorphism of vitamin $B_{12}$-binding (R) proteins of human saliva detected by isoelectric focusing. Biochem Genet 1979;17:909–920.

116 Owerbach D, Hjorth JP: Inheritance of a parotid secretory protein in mice and its use in determining salivary amylase quantitative variants. Genetics 1980;95:129–141.

117 Poulsen K, Jakobsen BK, Mikkelsen BM, et al: Coordination of murine parotid secretory protein and salivary amylase expression. EMBO J 1986;5:1891–1896.

118 Dlouhy SR, Taylor BA, Karn RC: The genes for mouse salivary androgen-binding protein (ABP) subunits alpha and gamma are located on chromosome 7. Genetics 1987;115:535–543.

119 Dlouhy SR, Nichols WC, Karn RC: Production of an antibody to mouse salivary androgen-binding protein (ABP) and its use in identifying a prostate protein produced by a gene distinct from Abp. Biochem Genet 1986;24:443–463.

120 Parker M, Needham M, White R: Prostatic steroid binding protein: gene duplication and steroid binding. Nature 1982;298:92–94.

121 Karn RC, Dlouhy SR, Springer KR, et al: Sex-limited genetic variation in a mouse salivary protein. Biochem Genet 1982;20:493–504.

122 Karn RC: Sex-limited effects of the expression of the *db* gene in mice during puberty. Biochem Genet 1981;19:355–371.

123 Karn RC, Dlouhy SR: Origins of variation of mouse salivary androgen-binding protein (ABP): An analysis of variants in the genus Mus and Comparison with other rodents (submitted, 1990).

124 Bing J, Poulsen K, Hackenthal E, et al: Renin in the submaxillary gland. A review. J Histochem Cytochem 1980;28:874–880.

125 Ekker M, Tronik D, Rougeon F: Extra-renal transcription of the renin genes in multiple tissues of mice and rats. Proc Natl Acad Sci USA 1989;86:5155–5158.

126 Wilson CM, Kimberlin DF, Griffin JE, et al: Specificity of androgen resistance in *Mus caroli* kidney. Biochem Genet 1988;26:705–716.
127 Maniatis T, Goodbourn S, Fischer JA: Regulation of inducible and tissue-specific gene expression. Science 1987;236:1237–1245.
128 Kim H-S, Maeda N: Structures of two HaeIII-type genes in the human salivary proline-rich protein multi-gene family. J Biol Chem 1986;261:6712–6718.
129 Wright PS, Carlson DM: Regulation of proline-rich and alpha amylase genes in parotid-hepatoma hybrid cells. FASEB J 1988;2:3104–3107.

Robert C. Karn, PhD, Department of Biological Sciences, Butler University, Indianapolis, IN 46208 (USA)

# Subject Index